Mammalian
Neuroendocrinology

B. T. Donovan

*Reader in Neuroendocrinology, Institute of Psychiatry,
University of London*

McGRAW-HILL

London · New York · Sydney · Toronto
Mexico · Johannesburg · Panama · Singapore

Published by
McGRAW-HILL Publishing Company Limited
MAIDENHEAD · BERKSHIRE · ENGLAND

07 094135 1

PRINTED AND BOUND IN GREAT BRITAIN

A title in the European Animal Biology Series

Consulting Editor

D. L. LEE
*Fellow of Christ's College,
University of Cambridge*

To Jean

Preface

The multidisciplinary subject matter of this book has anatomical, clinical, physiological, zoological and even sociological implications, since such topics as the control of fertility and some problems arising from excessively high population densities are encompassed. Inevitably, much of the information relevant to this theme is widely scattered in the literature and few general surveys in less than treatise form are available. For this reason, though ambitious in intent, the present volume has been kept to a modest size, for it is designed to be easily read, to bridge the gap between textbook and journal and to assist in the evaluation of research in this field, while giving some impression of controversial issues. Preference in the citation of papers has been given to recent publications, which, in turn, refer to earlier studies.

Help in the preparation of this work has come from many colleagues; in particular, those who have allowed me to reproduce illustrations from their publications, and those who have generously read some of the typescript. The source of borrowed figures is indicated in the legends, but to name my critical readers might prove unwise, for their good advice has not always been followed and they are not to blame for any errors of judgement.

Contents

1

Introduction

Neuroendocrinology deals with the interaction of the nervous and endocrine systems in the control of bodily function and in the adaptation of an individual to its environment. Far back in history it was believed that 'humors' determined emotional states and modified behaviour. Such humors were believed to be transported along nerves to the periphery, until the discovery of bioelectricity by Galvani and the rise of modern neurophysiology supplanted this concept. Chemical mediators have since been found to be vital in the transmission of the nervous impulse from nerve to muscle, and, at a greater distance, adrenaline and noradrenaline liberated in autonomic ganglia or from the adrenal medulla can exert effects far from the site of origin. The action of these chemical mediators lasts for but a short time and with the disappearance or destruction of the agent the *status quo* is restored. Another group of chemical substances

1

exerts more prolonged effects and the study of the action, formation, secretion, and control of release of these latter materials, hormones, falls within the scope of endocrinology.

Intensive study of the relationship between the nervous and endocrine systems in mammals is of recent date and has been largely directed toward an analysis of the factors involved in the neural control of the pituitary gland, although a variety of other functions, such as the physiology of the adrenal medulla, fall in the realm of the subject. The modern neuroendocrinologist is also concerned, not only with the neural control of the endocrine system, but also with the endocrine control of nervous activity, as exemplified by the behaviour and the attitudes to environmental stimuli of every individual. In fact, the mass of relevant knowledge is now so great that the present discourse will be limited mainly to the mammalian situation in order to facilitate a detailed discussion of neuroendocrine processes within a restricted compass. Zoologists are aware of innumerable instances in the invertebrates where development, growth, and reproduction are controlled by neurosecretory cells located within the nervous system. In crustaceans, for example, gonadal development is restrained by the nervous system, with a gonad-inhibiting hormone being produced by neurosecretory cells in the eyestalks of crabs. Moulting in insects is similarly controlled by the brain through a hormone produced by neurosecretory cells within the protocerebrum and released from storage in the corpora cardiaca. Many such instances of neuroendocrine interaction in invertebrates, fish, and amphibia are given by Scharrer and Scharrer (1963) in their thoughtful monograph on neuroendocrinology.

Meaningful study of brain–endocrine relationships could not begin until the foundations of endocrinology were laid at the beginning of this century. Although information indicative of the existence of an internal secretion was available from the work of Berthold in 1849 on testicular transplantation in the cock, it fell to Bayliss and Starling to propose the name 'hormone' in 1902 in the course of their studies of the part played by the duodenal hormone, secretin, in the control of pancreatic secretion. At about the same time the immediate and short-lived actions of extracts of the pituitary gland on the cardiovascular system, uterus and mammary glands were being investigated pharmacologically and related to the neurohypophysis, although as late as 1908 leading workers in this field considered that the anterior lobe of the pituitary gland had no physiological function. This misconception was later corrected on the basis of investigations involving surgical removal of the hypophysis but the consequences proved so variable that controversy raged until Philip Smith developed a satisfactory technique in the rat which avoided damage to the hypothalamus. He was able to show in 1927 that loss of the pituitary gland retarded growth and caused atrophy of the gonads, thyroid gland, adrenal cortex, liver, spleen, and kidneys. The observation that these changes could be reversed by the implantation of pituitary tissue, or by the administration of crude extracts, provided the starting point of much subsequent work on the separation and purification of pituitary hormones.

Indications that the nervous system exercises over-riding control over reproductive function were available long before the part played by the pituitary gland was understood. It seems that the earliest observation in this field was that of Martin Lister in 1675 who wrote in a letter to John Ray that swallows may be induced to lay more eggs than usual by daily removal of an egg from the nest. Later, in 1797, Haighton described the changes in the ovaries of the rabbit that reflexly follow the act of coitus, and the process of ovulation was studied in detail by Heape in 1905. Even so, a further thirty-one years elapsed before the demonstration that ovulation in the rabbit and pseudopregnancy in the rat could be induced by electrical stimuli applied to the head. Alongside work on the neural control of ovulation in the rabbit, numerous observations concerning the effect of emotional stimuli on reproductive rhythms were being collected and collated by Marshall in Cambridge and these, when published in 1936, led to the conclusion that the many external influences capable of modifying the sexual cycle act upon the pituitary gland through the central nervous system. In turn, Harris presented a detailed analysis of the factors concerned in the neural control of the pituitary gland in 1948. By that time, the close relationship between the hypothalamus and the control of the secretion of the neurohypophysial hormones through the nerve tracts in the pituitary stalk had been well studied and the principles underlying the function of the system established (chapter 6). For the adenohypophysis the situation was somewhat different. Although there was much evidence that the pars distalis was under neural control the identity of the connecting link was proving controversial.

Evidence that the secretion of anterior pituitary hormones is governed through the nervous system now is overwhelming. Light is a potent influence in the control of the timing of the breeding season in many species and acts through the visual system of the brain. The sight of another rabbit may provoke ovulation in the doe, while the introduction of a mirror into the cage of an isolated pigeon can provide sufficient stimulus for egg-laying. Tactile or proprioceptive stimuli may also be important, for, as in the case of the swallow referred to above, removal of eggs from the nest of certain birds causes laying to be resumed to make up the normal number. In mammals suckling favours the maintenance of lactation by promoting the secretion of pituitary lactogenic hormone. The sense of smell can also be important, although it is more concerned with the triggering of sexual behaviour. In mice the scent of a strange male can prevent the pregnancy expected from a fertile mating occurring a few hours earlier. Clinically, it is well known that the menstrual rhythm is readily altered by environmental stress such as occasioned by an unhappy love affair or the first weeks of nursing in a teaching hospital. Some pathological lesions in the brain can interfere with sexual development and function, but in children others may cause precocious puberty. Emotional stress also evokes the release of adrenocorticotrophic hormone from the pituitary gland as shown by an increased excretion of adrenal hormones in the urine, and can inhibit the output of thyrotrophic hormone.

With the realization that the activities of the pars distalis lay under neural control the search for the connecting link began. It seemed clear that the controlling influence traversed the pituitary stalk and it was likely that the anterior pituitary gland, like the posterior lobe, was innervated from the hypothalamus. Much effort was devoted to the identification of the nerve fibres involved and though positive results were reported these never met with universal acceptance, largely because the fibres described were few and inconsistent in appearance. Dispute over the existence of a secretomotor innervation of the pars distalis died down when the gland was examined with the electron microscope and nerve fibres could not be found in significant number. On the other hand, reticular connective tissue fibres are present and as it is difficult to distinguish between reticular and unmyelinated nerve fibres

3

with light microscopy the controversial findings made in the early investigations can be satisfactorily explained.

In reality, the functional link between the brain and the pituitary gland is provided by the hypophysial portal vessels, which were known to exist long before their significance became apparent. The evidence that they might form the anatomical path between the hypothalamus and pars distalis was summarized by Green and Harris in 1947. Experiments directed toward the analysis of the importance of the portal vessels were then undertaken and it was quickly discovered that the return of function after transection of the pituitary stalk could be closely correlated with regeneration of the vessels and the bridging of the break in the pituitary stalk. When regeneration of the portal vessels was prevented by the insertion of a barrier between the cut ends of the stalk gonadal atrophy and other indications of pituitary hypofunction became apparent. That blood reaching the pituitary gland through the portal vessels possessed some special properties from the point of view of pituitary function was indicated by experiments in which pituitary tissue was transplanted to a kidney or to another part of the brain away from the hypothalamus in an hypophysectomized animal, when normal function was not resumed although the graft obtained a fresh supply of blood from the host organ. Only grafts applied to the median eminence of the hypothalamus within reach of the portal vessels secreted hormones in the usual manner. Observations of this kind form the basis of the neurohumoral theory of the control of anterior pituitary function which argues that nerve fibres in the hypothalamus liberate chemical agents into the portal vessels which are carried down the pituitary stalk to modify the activity of the cells in the pars distalis. Substantial confirmation of this idea has come from study of the effects of extracts of hypothalamic tissue on pituitary function and the demonstration that the output of follicle-stimulating hormone, luteinizing hormone, adrenocorticotrophic hormone, thyrotrophic hormone, and growth hormone can be increased, and that of prolactin inhibited, by the application of the appropriate extract. Much work has been done on the isolation and purification of the so-called 'releasing factors' and this will be discussed later.

As a corrective to the attention given to the neurohumoral control of anterior pituitary function in recent years it should be pointed out that other neuroendocrine phenomena have a longer history. The medulla of the adrenal gland secretes two hormones, adrenaline and noradrenaline, whose effects mimic those produced by excitation of the sympathetic system. Study of this gland by Cannon and his associates in the 1920s led to the demonstration that the two sympathetic amines were released in association with emotional activity or in response to stress. Cannon believed that the adrenal medulla, which receives a secretomotor innervation from the sympathetic system, was activated whenever a fight or flight seemed necessary, whether or not physical activity resulted. His views were set out in a famous book on *Bodily Changes in Pain, Hunger, Fear and Rage* which was published in 1929. Consideration of the consistent changes in the cortex of the adrenal gland which occur in response to potentially damaging agencies led to the later development of the 'stress' concept by Selye which held the attention of many investigators in the 1950s.

The posterior lobe of the hypophysis is connected to two prominent groups of cells in the hypothalamus, the supraoptic and paraventricular nuclei, by a well defined tract of nerve fibres running through the pituitary stalk. This system forms part of the neurohypophysis, which elaborates polypeptides concerned with the control of water excretion, uterine activity, and milk ejection; diabetes insipidus is a characteristic feature of cases of hypofunction of this system. A connection between diabetes insipidus and the neuro-hypophysis began to emerge around 1912 when Crowe, Cushing, and Homans reported that transection of the pituitary stalk in dogs caused severe but transient polyuria, and Frank described a case of polyuria in a man who had suffered a bullet wound in the head involving the sella turcica. The later experimental work of Fisher, Ingram, and Ranson (1938) which demonstrated that permanent polyuria was produced by destruction of the tract between the supraoptic nuclei and the posterior lobe by localized lesions in the hypo-thalamus closed the pioneer era in the physiology of the neurohypophysis. The realization that the cells of the supraoptic and paraventricular nuclei were neurosecretory and produced hormones acting on the kidneys, uterus and mammary glands, still lay in the future (chapter 5).

From the examples of neuroendocrine function so far presented it is possible to discern several ways in which the nervous system can influence bodily function by humoral means (Figure 1.1). There can be transmission of a

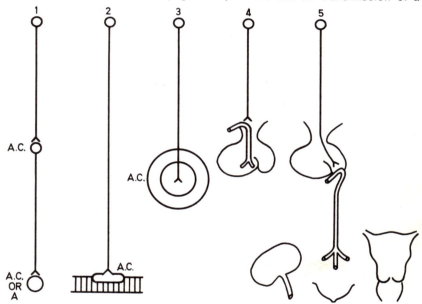

FIGURE 1.1 To compare different systems in which humoral transmission of stimuli may occur: *1*, autonomic nervous system; *2*, neuromuscular ending; *3*, sympatho-adrenal medullary ending. In these three systems it is established that a cholinergic (A.C.) or adrenergic (A) substance is liberated from the nerve terminal and acts directly on the effector cell. *4*, Hypothalamo-adenohypophysial system in which a short vascular pathway intervenes between the nerve terminal and effector cell located in the pars distalis. *5*, Hypothalamo-neurohypophysial system in which a long vascular pathway (the systemic circulation) intervenes between the nerve terminal in the neural lobe and the effector cells in the kidney, mammary gland, or uterus. From Harris (1960).

stimulus directly from nerve to muscle across a minute gap at a motor end plate by a chemical process that requires acetylcholine. Slightly more complicated is the sympathetic pathway which involves two synapses, one employing the short-lived acetylcholine and a second at which an adrenergic substance with a longer life is involved. Alternatively, as illustrated by the adrenal medulla, the adrenergic synapse is dispensed with and the sympathetic mediator liberated into the general circulation so that it can exert widespread effects. Neuronal products of a different kind are directly involved in another variant of the process, whereby the neurohypophysial hormones are released from the posterior lobe of the pituitary gland into the general circulation. These hormones, like those of the adrenal medulla, are discharged in response to a barrage of cholinergic impulses applied to the cells of origin within the hypothalamus. Finally, humoral mediators concerned with anterior pituitary function (to be discussed in detail later) are released into the hypophysial portal vessels to traverse a short and well defined humoral pathway to promote the secretion of hormones by the cells of the pars distalis. The pituitary hormones then pass into the systemic circulation to act upon a wide range of end organs. The humoral mediators have been known as 'releasing factors' almost since their discovery but certain shortcomings in this term are becoming apparent. When acting upon the pituitary gland the factors do more than just cause the release of a particular hormone: synthesis also occurs, though this may follow secondarily as a consequence of the discharge of hormone. In the case of two pituitary hormones, prolactin and melanocyte-stimulating hormone, the hypothalamic humoral agents act to inhibit release of the hormone from the hypophysis, and here the term 'releasing factor' is clearly inappropriate and has been replaced by 'inhibiting factor'. With these considerations in mind Schally, Arimura, Bowers, Kastin, Sawano, and Redding (1968) have urged that the hypothalamic releasing factors should be regarded as hormones. In order to exclude toxic materials from consideration the true hormone would be capable of releasing a given pituitary hormone by a direct action on the gland cells, and its action should not involve damage to the tissue. Criteria which should be employed in the evaluation of the activity of any proposed neurohumoral agent are discussed on p. 72. The suggested nomenclature is set out in Table 1.1, alongside the present system, from which it will be evident that complications exist in both. The cumbersome 'melanocyte-stimulating hormone release-inhibiting factor' of the present system has its counterpart in the new version.

Endocrine control mechanisms can be examined theoretically in another way with the possible interactions being set out in hierarchical form as in Figure 1.2, which comes from the work of Scharrer and Scharrer (1963). No one endocrine function uses all of the connections but there is evidence for each of the associations depicted by an arrow, though not always in mammals. Thus, the central nervous system produces oxytocin which acts on the mammary gland (a target organ) to cause milk ejection, while stimuli from the nipple of the mammary gland, initiated by suckling, can provoke the release of the hormone. An endocrine organ B (ovary) can be influenced indirectly through endocrine organ A (the hypophysis) or directly, as in crustaceans. The secretion of adrenaline and noradrenaline from the adrenal medulla (an endocrine organ B) is directly controlled by the brain through the splanchnic nerves. Growth of the mammary gland (a target organ) is immediately influenced by prolactin from the pituitary gland (endocrine organ A), as well as by the sex hormones from the ovaries (endocrine organs B). And the secretion of pro-lactin is governed by the brain (chapter 11). Limitations on the secretion of hormones are provided by a feedback action of the target organ hormone upon the gland supplying the trophic stimulus, or its hierarchical superior, or by neural stimuli from the target organs. In some mammals (guinea-pig, sheep) the functional life of the corpora lutea in the ovaries is determined by a sub-stance produced by the uterus (a target organ) and further, sex hormones produced by the ovaries modify the secretion of gonadotrophin by the hypophysis.

TABLE 1.1   THE NOMENCLATURE OF SOME HYPOTHALAMIC NEUROHUMORS*

Present Name Hypothalamic Factor	Abbreviation	Proposed Name Hypothalamic Hormone	Abbreviation
Corticotrophin-releasing factor	CRF	Corticotrophin-releasing hormone	CRH
Luteinizing hormone-releasing factor	LRF or LH-RF	Luteinizing hormone-releasing hormone	LH-RH or LRH
Follicle-stimulating hormone-releasing factor	FSH-RF	Follicle-stimulating hormone-releasing hormone	FSH-RH or FRH
Thyrotrophin-releasing factor	TRF	Thyrotrophin-releasing hormone	TRH
Growth hormone-releasing factor or somatotrophin releasing factor	GRF or SRF	Growth hormone-releasing hormone or somatotrophin-releasing hormone	GRH or SRH
Prolactin-inhibiting factor (mammals)	PIF	Prolactin-release-inhibiting hormone	PRIH
Melanocyte-stimulating hormone (MSH) release-inhibiting factor	MIF	MSH-release-inhibiting hormone	MRIH

* From Schally, Arimura, Bowers, Kastin, Sawano, and Redding, 1968

The feedback action of target organ hormones upon the brain figures largely in most considerations of neuroendocrine function and may be positive or negative in character. Negative feedback is most familiar, for here an excess of the target organ hormone acts back on the controlling mechanism at a brain–pituitary level to suppress the secretion of excessive amounts of the appropriate trophic hormone. Left to itself such a control process would produce an oscillation of the blood level of both trophic and target organ hormone about a preset optimal level, but physiologically the feedback action of target organ hormones can often be overridden by a neural drive acting through the hypothalamus. On occasion a target organ hormone can promote the secretion of a trophic hormone. Suitably timed administration of the ovarian hormone progesterone can elicit the release of the gonadotrophin causing follicular rupture and so trigger ovulation. In this case the ovarian steroid is exerting a positive feedback action. One of the features of modern

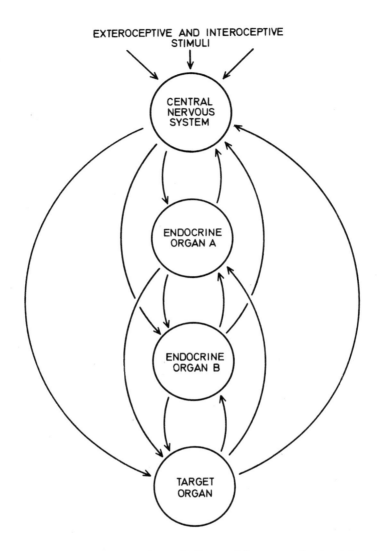

EXTEROCEPTIVE AND INTEROCEPTIVE
STIMULI

CENTRAL
NERVOUS
SYSTEM

ENDOCRINE
ORGAN A

ENDOCRINE
ORGAN B

TARGET
ORGAN

FIGURE 1.2 Diagram illustrating theoretically possible neuroendocrine interactions. The arrows indicate both nervous and hormonal connections, but not all pathways are employed in any particular species. From Scharrer and Scharrer (1963).

endocrinology has been the shift in the focus of feedback action from the pituitary gland to the hypothalamus. While this is a closer approximation of the truth it is known that the activities of the pars distalis can be directly affected by target organ hormones, as will emerge later on. It has also become apparent that the pituitary hormones can themselves act directly upon the hypothalamus to influence their own secretion. This direct feedback action

8

can also cause certain patterns of behaviour to become manifest, such as when prolactin induces maternal behaviour in doves.

Despite the mass of material dealing with the action of hormones on the brain very little is known of the ways in which these effects are brought about. Understanding of the manner in which the activities of the hypothalamus is subject to control by the higher centres is also very poor. Several functional circuits within the brain, notably the olfactory and limbic systems, exert a peculiar influence over the secretion of hormones by the pituitary gland but the information available remains fragmentary. This may be due in part to the difficulties that arise in experiments aimed at the destruction or stimulation of precise, but diffuse, areas of the brain with minimal damage to neighbouring structures. The fact that such studies tend to fall in the no-man's land between physiology and psychology is probably of no less significance.

2

The Foundations of
Neuroendocrine Mechanisms

Although almost any afferent nerve ending can supply information of significance in the neural control of endocrine function, certain parts of the nervous
system are of special importance. These are largely confined to the subcortical parts of the brain and to the phylogenetically older parts of the
cerebral hemispheres, for in most mammals the most recently developed
neocortex, which is responsible for the intellectual development of man, is of
less significance. This is not to say that the neocortex lacks influence over
neuroendocrine function, but rather to emphasize that it acts only to modulate
the more fundamental mechanism provided by brain-stem structures. Accordingly, most of the information on the role of the neocortex in the control of the
endocrine system comes from work on primates and in man.

Developmentally, the brain is tubular in structure and three primary divisions can be distinguished: the forebrain, midbrain, and hindbrain. The forebrain forms the anterior bulb of the neural tube and two lateral outgrowths from it provide the presumptive cerebral hemispheres. The part of the forebrain that remains, and comes to lie between the two hemispheres, differentiates into the thalamus, epithalamus, and hypothalamus, or diencephalon. The hypothalamus occupies, as it were, the floor of the diencephalon, where its limits are fairly well marked by the arteries of the Circle of Willis, although these also enclose the optic chiasma. The pituitary stalk is attached to the centre of a protuberance on the surface of the hypothalamus, the tuber cinereum. Anterior to the thalamus and hypothalamus lies the septum (Figure 2.7). This is a mass of grey matter that merges with the medial walls of both hemispheres and is prominent in lower mammals. With the extensive development of the neocortex and of a large commissure between the two cerebral hemispheres in primates it becomes stretched and thinned. The septum is connected to the amygdaloid nuclei, hippocampus and hypothalamus, by somewhat complex fibre systems (de Groot, 1966a).

In the course of evolution, differential growth of part of each cerebral hemisphere has dwarfed that of the rest, so that the primitive cortex becomes displaced medially and forms a border, or limbus, around the diencephalon. This is referred to as the limbic lobe. Basally, part of the cortex becomes rolled inward, is completely invested by neocortex, and forms the hippocampus (Figures 2.6 and 2.7). The limbic lobe and hippocampus are major contributors to the limbic system, which includes the septum, hypothalamus, and certain other structures such as the amygdala, and is assuming ever greater prominence in neuroendocrine study as more becomes known about the control of endocrine and behavioural processes. The amygdala, or complex of amygdaloid nuclei, are located in the basal region of each temporal lobe (Figures 2.3, 2.4, 2.7, and 2.8) and can be regarded as positioned at the crossroads between the ancient and more recently differentiated cortex, between the olfactory system, hippocampus and basal ganglia, and between cortex, septum complex, and hypothalamus (de Groot, 1966a).

Of all the structures mentioned above, the hypothalamus, with its coupling to the pituitary gland through the hypophysial portal vessels and the nerve tracts in the pituitary stalk, has proved to date to be of greatest interest to the neuroendocrinologist. However, the activities of the hypothalamus are influenced in turn by other parts of the brain, such as the amygdaloid region of the temporal lobe, the hippocampus, and the visual and auditory systems, as well as by psychic stimuli through the neocortex. In view of these interactions it is advisable to review certain pertinent neuroanatomical relationships in a little more detail and to introduce some physiological matters.

The Hypothalamus

Since so much attention has been paid to the functions of the hypothalamus, and as constant reference will be made to certain of its components, it is as well to sketch some of the anatomical features of this region. These are illustrated by Figures 2.1, 2.2, 2.3, and 2.4, which are based on a standard atlas of the rat brain prepared by de Groot (1959) and used in many research laboratories. A sagittal section is presented as Figure 2.1, and transverse sections taken at the points indicated by the arrows provided as Figures 2.2,

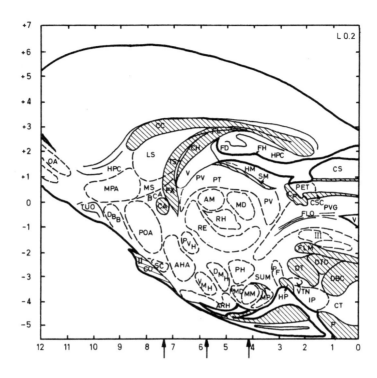

FIGURE 2.1 The location of some neural structures in the brain of the rat, plotted on a sagittal section. The levels at which the transverse sections of the brain shown in Figures 2.2, 2.3, and 2.4 are taken are indicated by the arrows and it should be noted that the nuclei do not lie in the midline. From de Groot (1959).

The abbreviations stand for:

A A A	Area amygdaloidea anterior
A B L	Nucleus amygdaloideus basalis pars lateralis
A C B	Nucleus accumbens septi (area parolfactoria lateralis)
A C E	Nucleus amygdaloideus centralis
A C O	Nucleus amygdaloideus corticalis
A D	Nucleus anterodorsalis thalami
A H A	Area anterior hypothalami
A L	Nucleus amygdaloideus lateralis
A M	Nucleus anteromedialis thalami
A ME	Nucleus amygdaloideus medialis
A R H	Nucleus arcuatus hypothalami
A V	Nucleus anteroventralis thalami
B C A	Nucleus proprius commissurae anterioris (Bed nucleus)
B S T	Nucleus proprius striae terminalis (Bed nucleus)
C A	Commissura anterior
C C	Corpus callosum

12

CE	Capsula externa
CH	Commissura hippocampi
CI	Capsula interna
CLA	Claustrum
CO	Chiasma opticum
CP	Commissura posterior
CPU	Nucleus caudatus/Putamen
CS	Colliculus superior
CSC	Commissura colliculi superioris
CT	Nucleus centralis tegmenti (Bechterew)
DBB	Gyrus diagonalis (Diagonal Band of Broca)
DBC	Decussatio brachiorum conjunctivorum
DMH	Nucleus dorsomedialis hypothalami
DTD	Decussatio tegmenti dorsalis (Meynert)
DTV	Decussatio tegmenti ventralis (Forel)
EP	Nucleus entopeduncularis
FA	Fissura amygdaloidea
FD	Gyrus dentatus (Fascia dentata)
FH	Fissura hippocampi
FI	Fimbria hippocampi
FL	Fornix longus
FLD	Fasciculus longitudinalis dorsalis (Schütz)
FLM	Fasciculus longitudinalis medialis
FR	Fissura rhinalis
FX	Fornix
GP	Globus pallidus
HL	Nucleus habenularis lateralis
HM	Nucleus habenularis medialis
HP	Tractus habenulo-interpeduncularis
HPC	Hippocampus
ICL	Nucleus amygdaloideus intercalatus
IP	Nucleus interpeduncularis
LHA	Area lateralis hypothalami
LS	Nucleus lateralis septi
LT	Nucleus lateralis thalami
MD	Nucleus mediodorsalis thalami
MFB	Fasciculus medialis telencephali (Medial forebrain bundle)
ML	Nucleus mamillaris lateralis
MM	Nucleus mamillaris medialis
MP	Nucleus mamillaris posterior
MPA	Area parolfactoria medialis
MS	Nucleus medialis septi
MT	Tractus mamillo-thalamicus (Vicq d'Azyr)
NCP	Nucleus proprius commissurae posterioris (Bed nucleus)
NOT	Nucleus tractus olfactorius
OA	Nucleus olfactorius anterior
OT	Tractus opticus
P	Pons
PC	Peduncularis cerebri
PF	Nucleus parafascicularis thalami
PH	Nucleus posterior hypothalami
PIR	Cortex piriformis
PM	Pedunculus mamillaris
PMD	Nucleus premamillaris dorsalis
POA	Area preoptica (medialis, lateralis)
PRT	Area pretectalis
PT	Nucleus parataenialis thalami
PV	Nucleus paraventricularis thalami
PVG	Substantia grisea periventricularis
PVH	Nucleus paraventricularis hypothalami
RE	Nucleus reuniens thalami
RH	Nucleus rhomboideus thalami
RT	Nucleus reticularis thalami
S	Subiculum
SC	Nucleus suprachiasmaticus
SM	Stria medullaris thalami
SO	Nucleus supraopticus hypothalami
ST	Stria terminalis
SUM	Area supramamillaris
TOL	Tractus olfactorius lateralis
TS	Nucleus triangularis septi
TT	Tractus mamillo-tegmentalis
TUO	Tuberculum olfactorium
TZ	Zona transitionalis
V	Ventriculus cerebri
VA	Nucleus ventralis thalami, pars anterior
VD	Nucleus ventralis thalami, pars dorsomedialis
VE	Nucleus ventralis thalami
VM	Nucleus ventralis thalami, pars medialis
VMH	Nucleus ventromedialis hypothalami
VTN	Nucleus ventralis tegmenti (Tsai)
ZI	Zona incerta
II	Nervus opticus
III	Nucleus nervi oculomotorii
V	Nucleus tractus mesencephalici nervi trigemini

2.3, and 2.4. The main components of the hypothalamus are to be found in all mammals, although some nuclei are more prominent in one species than in another. The human hypothalamus appears compressed and distorted when compared with that of the rat and other non-primate mammals.

Morphologically, the anterior border of the hypothalamus is marked by the rostral edge of the optic chiasma, and its caudal limit is coincident with the

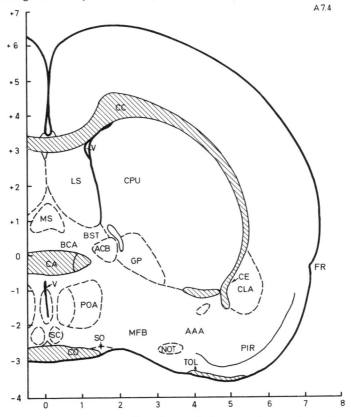

FIGURE 2.2 Transverse section of the brain of the rat, 7·4 mm anterior to ear zero (p. 12). From de Groot (1959).
Abbreviations as in Figure 2.1.

caudal tips of the mamillary bodies. The lateral boundaries of the hypo- thalamus are largely formed by the optic tracts, while dorsally an inconspicuous indentation in the walls of the third ventricle (which longitudinally splits the hypothalamus into two halves) marks the boundary with the thalamus.

The optic chiasma, tuber cinereum, median eminence, pituitary stalk, and mamillary bodies are prominent features of the ventral surface of this part of the brain.

The hypothalamus can be divided into three longitudinal zones (Crosby, Humphrey, and Lauer, 1962). These are the periventricular zone, which is

14

close to the third ventricle and contains many fine medullated and un-medullated fibres running obliquely in a dorsoventral direction, the medial zone, which contains a number of important cell masses, and a lateral zone, which is made up largely of nerve fibres passing through the hypothalamus in a longitudinal direction.

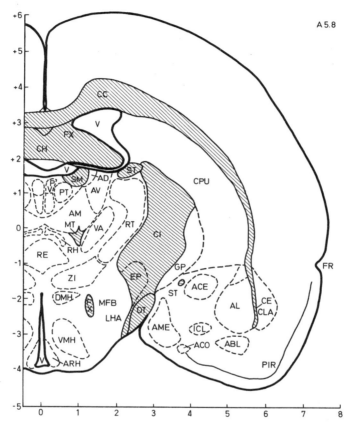

FIGURE 2.3 Transverse section of the brain of the rat, 5·8 mm anterior to ear zero (p. 12). From de Groot (1959).
Abbreviations as in Figure 2.1.

Histologically, it is difficult to separate the preoptic area from the anterior hypothalamus, for the two zones merge. As their name suggests the supra-chiasmatic nuclei are located above the optic chiasma near the midline, while more laterally the supraoptic nuclei straddle the lateral parts of the optic chiasma when observed in frontal sections. The large cells of the supra-optic nuclei are very conspicuous in histological sections and possess peculiar staining properties due to their content of neurosecretory material. Similar large cells are to be found in two groups dorsocaudally on each side of the third ventricle and these comprise the paraventricular or filiform nuclei.

The axons of both the supraoptic and paraventricular nuclei run toward the pituitary stalk to enter the infundibular process and the neurosecretory material travels down the axons to be liberated from that body. Since its discovery the phenomenon of neurosecretion has aroused so much curiosity and interest that the study of various facets of the production, transport, release, and function of the material has developed into a sub-speciality of its own. Neurosecretion will be described in more detail in chapter 5. The arcuate nuclei lie

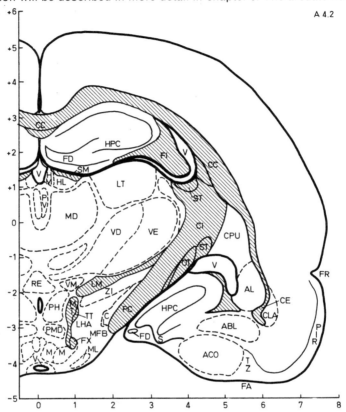

FIGURE 2.4 Transverse section of the brain of the rat, 4·2 mm anterior to ear zero (p. 12). From de Groot (1959).
Abbreviations as in Figure 2.1.

in the ventral part of the tuberal region in the periventricular grey matter, and have often been associated with the control of gonadotrophin secretion.

The ventromedial nuclei lie below and behind the paraventricular complexes. These are round or oval complex nuclear masses which are clearly defined in laboratory rodents, and seem to be involved in a variety of activities, including the control of food intake, the expression of emotions and the secretion of pituitary hormones. Two types of neurons have been distinguished in the ventromedial nuclei of the cat: a bipolar form which predominates at the lateral border, and a multipolar type which is most evident in the ventral portions of the nuclei. It has been suggested that the bipolar cells may be interneurones on inhibitory pathways from the amygdala (Murphy and Renaud, 1969). Above and slightly rostral to the ventromedial nuclei lie the dorsomedial nuclei.

The mamillary bodies are prominent components of the hypothalamus and are relatively large in man when compared with other species. They can be divided into large medial and small lateral nuclei but the precise nature of the lateral nuclei is not well understood, since the name has been applied to several cell groups in the area of the mamillary bodies. Prominent connections between the mamillary bodies and thalamus (mamillo-thalamic tract, bundle of Vicq d'Azyr), hippocampus, tegmentum, and interpeduncular nuclei exist, but rather little is known of the function of the mamillary bodies, although the mamillo-thalamic tract is an important constituent of the limbic circuit of Papez (p. 25).

Afferent Connections of the Hypothalamus

Many parts of the brain contribute to the influx of information to the hypothalamus. These are illustrated in diagrammatic form in Figures 2.5 and 2.6, which come from the work of Raisman (1966, 1969). The important structures in this regard are the septum, hippocampus, anterior thalamus, amygdala, piriform cortex, and midbrain (Figure 2.5), although additional areas influence hypothalamic activity indirectly. As will emerge later, the septum, hippocampus, amygdala, and piriform cortex are involved in the control of adrenal, thyroid, and gonadal function, while stimuli from the body surface and visceral organs are transmitted to the hypothalamus through the midbrain. In view of the barrage of afferent impulses from so many other regions of the brain (Figure 2.5), it is evident that a high degree of integration and differentiation must occur. Perhaps convergence of information is indicated by a surprising lack of specificity on the part of cellular units in the hypothalamus, for the activity of a single unit can be influenced in a variety of ways (chapter 16).

A major source of afferent input reaches the hypothalamus through the medial forebrain bundle. Despite its name, the medial forebrain bundle is a meshwork of unmyelinated fibres and it is difficult to trace distinct pathways through it because it is composed of many short relays. Both afferent fibres to and projections from the hypothalamus travel in the medial forebrain bundle, which extends rostrally through the preoptic area to the olfactory region and caudally to the midbrain. Raisman (1966) considers that most of the extrinsic connections of the hypothalamus are made with the medial forebrain bundle rather than with the medial hypothalamic nuclei. Thus the bundle is connected with the olfactory bulbs and tracts and the olfactory and medial orbitofrontal cortex, as well as with the septum, amygdala, and caudate nuclei, while its relationships with the hypothalamic nuclei must not be overlooked (Figure 2.6). Although general agreement has not been reached over the question of the existence of direct neocortical-hypothalamic fibres in the medial forebrain bundle there is little doubt that indirect pathways through the septum and thalamus exist.

The paired columns of the fornix pass through the anterior hypothalamus and are prominent landmarks in this area. Each fornix arises in the hippocampus of each temporal lobe of the brain and forms an arc which initially passes

17

medio-caudally, swings upwards over the diencephalon and then runs forward to curve down and caudal again in the neighbourhood of the anterior commissure. The fornices then part company and end mainly in the mamillary bodies. Connections between the fornices and the septal areas exist and, by branches around the anterior commissure, the preoptic and anterior hypothalamic areas, as well as the tuber cinereum, are also supplied. The main (postcommissural) branch of the fornix may also carry fibres to the anterior hypothalamic area and to the ventromedial nuclei. The fornix forms part of the limbic system which will be described later.

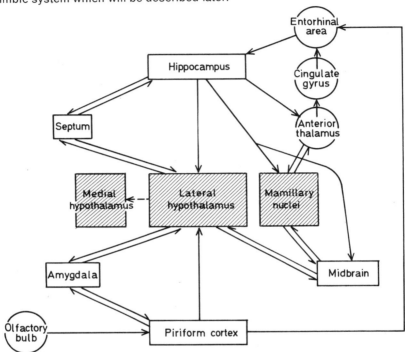

FIGURE 2.5 A schematic diagram of the principal fibre connections of the hypothalamus. From Raisman (1966).

Fibres from the amygdaloid nuclei in each temporal lobe come together to form the stria terminalis which makes a circuit around the diencephalon to enter the anterior hypothalamus in company with the fornix and medial forebrain bundle. The amygdaloid nuclei are involved in many endocrine and behavioural reactions and are connected with the preoptic area as well as the anterior hypothalamus, where links with the ventromedial and arcuate nuclei have been described. Other afferents to the hypothalamus from the amygdaloid nuclei adopt a horizontal course and travel medially over the optic tract to enter the lateral hypothalamus and medial forebrain bundle. Olfactory impulses that reach the corticomedial portion of the amygdaloid nuclei are transferred to the basolateral nuclei, from whence they pass to the hippocampus, fornix, and finally to the septal region, preoptic area and hypothalamus.

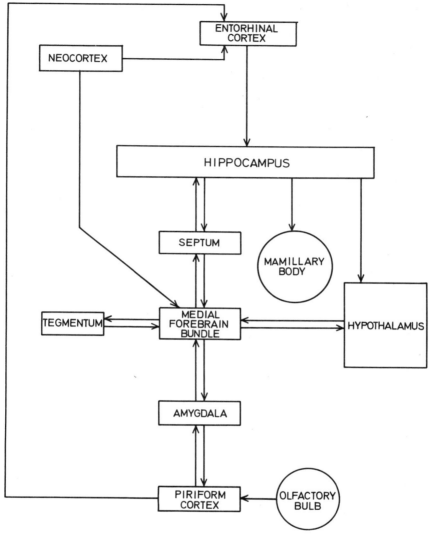

FIGURE 2.6 Some of the connections between the limbic system, medial forebrain bundle, and hypothalamus. After Raisman (1969).

Since the visual system plays such a large part in the control of neuro-endocrine function the possibility that fibres leave the optic chiasma to enter the hypothalamus has often been considered. Some authors have described such fibres while others have denied their presence. At the present time it is probably fair to state that convincing proof of the existence of such fibres has not been obtained. Impulses of visual origin do reach the hypothalamus but are relayed several times on their way to the hypothalamus. They could travel through the reticular formation or be passed on from the lateral geniculate bodies.

Numerous fibres run ventrally from the thalamus to the hypothalamus in the periventricular system and could convey information concerned with somatic and visceral sensation. Connections from the anterior nuclei of the thalamus to the medial mamillary nuclei through the mamillo-thalamic tracts have also been described.

Caudally the mamillary peduncle supplies fibres to the mamillary nuclei (chiefly the lateral) and possibly the tuberal region. This pathway may be fed by collaterals from the ascending sensory systems. Many connections between the hypothalamus and midbrain exist although they are difficult to trace anatomically. Fibres have been followed through the medial forebrain bundle to the lateral hypothalamic and preoptic nuclei as well as to the nucleus of the diagonal band of Broca and septal area. They form part of the input from the reticular formation to the hypothalamus which is of great physiological significance.

Efferent Connections of the Hypothalamus

The efferent connections of the hypothalamus are probably of more importance from the point of view of the control of behaviour and of autonomic function than from the strictly neuroendocrine standpoint, where the hypophysial portal vessels provide the operative link from the median eminence to the pituitary gland. However, the axons of the cells in the paraventricular and supraoptic nuclei, together with fibres from the tuberal region, make up tracts which converge upon the pituitary stalk to supply the infundibular process. It is down these axons that the neurosecretory products pass.

Fibres run forward to the septal, medial orbito-frontal, and parolfactory regions through the medial forebrain bundle, which thus conveys both afferent and efferent impulses (Figure 2.6), while there is a reciprocal projection from the hypothalamus to the amygdala along the pathways already described. Other nerve fibres run vertically close to the third ventricle to form a periventricular outflow to the thalamus, while yet others adopt a caudal route near to the midline to supply the dorsal longitudinal fasciculus which connects with numerous brain-stem centres and nuclei. The preoptic and anterior hypothalamic areas are two of the regions connected with the habenular nuclei through the stria medullaris.

From the medial mamillary nucleus a large bundle of fibres emerges dorsally and bifurcates, with one branch running backwards to the brain stem to make up the mamillo-tegmental tract which reaches the tegmental nuclei and connects with motor and preganglionic cell groups. The other branch passes rostro-dorsally to supply the anterior nuclear group of the thalamus. This forms the mamillo-thalamic tract.

The Functions of the Hypothalamus

Besides being of prime importance in the control of endocrine function the hypothalamus is concerned with many other vital activities. These will be briefly surveyed, if only to give some indication of the multiplicity of function displayed by this area and to show how difficult it is to study one aspect of hypothalamic physiology to the exclusion of others.

Olfactory and visceral information becomes integrated within the hypo-thalamus, which in turn discharges to centres concerned with the secretion of saliva and gastric juice, and with the motor nuclei concerned with chewing and swallowing. It is well known that lesions in the tuberal region of the hypo-thalamus produce adiposity, as well as savage behaviour, and the ventro-medial, dorsomedial, and arcuate nuclei figure large in the structures destroyed. On the other hand, damage to the lateral hypothalamus, on the outer side of the dorsomedial nucleus, causes loss of desire to eat or aphagia. It is on this basis that the existence of 'feeding' and 'satiety' centres in the hypo-thalamus is postulated, for stimulation of the lateral hypothalamic area (destruction of which causes aphagia) induces excessive eating on the part of the cats studied. The anterior and lateral areas of the hypothalamus are also important in the control of water balance, and the role of the hypothalamus in governing metabolic processes will be discussed in detail in chapter 15.

Autonomic function is profoundly influenced by the hypothalamus. At one time the anterior part of the hypothalamus was said to govern parasympathetic activity, while sympathetic processes lay under the control of the posterior hypothalamus. However, this subdivision of labour is probably a matter of emphasis only, for the preoptic area, the anterior hypothalamic and dorso-medial nuclei, the dorsal hypothalamic area, and the mamillary bodies are all said to connect with parasympathetic centres in the brain stem. The ventromedial hypothalamic nuclei, parts of the posterior hypothalamic area and the region dorsolateral to the mamillary bodies are particularly associated with sympathetic responses. Stimulation of the preoptic area causes dilation of cutaneous blood vessels, a fall in blood pressure and slowing of the heart. Conversely, stimulation of the posterior hypothalamic region increases the heart rate. Changes in gastrointestinal motility, or peristalsis, occur with stimulation of the hypothalamus.

Temperature regulation is a further process under the sway of the hypo-thalamus. Local heating of the hypothalamus or electrical stimulation of the preoptic or anterior hypothalamic areas produces sweating in man and panting in animals together with vasodilation. Damage to the same area impairs the heat loss mechanisms so that the body temperature rises to levels that can prove fatal. Shivering forms part of the protective mechanism against cold and it has been suggested that shivering is controlled through the caudal hypothalamus.

The hypothalamus acts as a relay in the expression of emotion, or rather the bodily accompaniments of emotion such as changes in facial expression, heart rate, blood pressure, sweating, piloerection, and so on. Removal of the cerebral hemispheres and the thalamus appears to relieve the hypothalamus of inhibition for in cats so treated sham rage readily follows mild disturbance. Sham rage is also known to follow stimulation of the caudal hypothalamus. Destructive lesions in the ventromedial hypothalamic nuclei or irritative lesions in the dorsomedial nuclei will convert a tame and friendly cat into a savage and intractable beast. With intermittent electrical stimulation of the hypothalamus in the cat a pattern of responses ensues which includes a rise in blood pressure, muscle vasodilation, pupillary dilation, retraction of the nictitating membranes, spitting, vocalization, and hissing. This complex is termed the 'defence reaction', for obvious reasons (Hilton, 1966).

There is also good evidence that the hypothalamus is involved in the opera-tion of the 24-hour clock or circadian rhythm that is apparent in many physiological functions. In the rat, for example, running activity, food intake, water intake, and reproductive cycles are readily followed under laboratory conditions. Under a sequence of alternating 12-hour periods of light and darkness Richter (1965) has shown that normal rats exhibit a fairly continuous active period during darkness. The time of onset of these periods of activity varies from individual to individual and they do not necessarily occur at exactly 24-hour intervals. In one group of 60 rats the intervals ranged from 41 minutes less than 24 hours to 28 minutes more than one day. However, the intervals are extraordinarily consistent for any one animal. It has proved difficult to stop the clock by such experimental interventions as blinding, hypophysectomy, adrenalectomy, pinealectomy, starvation, dehydration, hypothermia, or ablation of almost every part of the brain. Only damage to the hypothalamus in the region of the ventromedial nuclei has proved effective and this is followed by a more or less constant stuporous condition which is interrupted by frequent and rather regular intervals of eating and drinking.

The Limbic System

Current interest in the functions of the limbic system can be traced back to a theoretical explanation of the neuroanatomical basis of emotion advanced by Papez in 1937, although the limbic lobe, which includes the infolded hippocampus derived from the wall of the primitive cerebral hemisphere, received its name from Broca because it completely surrounds the hilus of the hemisphere (Figure 2.7). It can also be regarded as forming a boundary or limbus between the telencephalon and the brain-stem, but agreement has not yet been reached on the precise definition and enumeration of limbic structures. The main pathways to and from the limbic lobe are provided by the fornix, the medial forebrain bundle in conjunction with the stria medullaris and stria terminalis, and the anterior thalamocingulate projections, with the major outflow being carried through the fornix (White, 1965). Once considered to be mainly concerned with olfaction, the limbic system is now believed to be involved in many other processes, for stimulation or destruction of the amygdala and immediately adjacent structures in the temporal lobe can effect a variety of bodily functions.

The olfactory apparatus and pathways make up a prominent part of the limbic system with subcortical connections which include the amygdala, septal nuclei, and hypothalamus. The amygdaloid nuclei can be divided into several nuclear groups of which the corticomedial nuclei, considered to be phylogenetically older and more particularly concerned with olfaction, and the basolateral nuclei are particularly noteworthy. As remarked earlier, the amygdaloid nuclei provide an important junction between the olfactory system, hippocampus, and basal ganglia, and between the cortex, septum complex, and the hypothalamus (de Groot, 1966; Figure 2.8). The amygdaloid nuclei receive an input from all sensory modalities, and convergence of afferents handling several modalities on to a single amygdaloid cell has been observed electrophysiologically. In addition to connections with neighbouring regions of the temporal lobes links with the brain-stem reticular formation, cerebellum, hippocampus, thalamus, and hypothalamus through the stria terminalis are known. The primary projections of the amygdala connect with the basal septal region, the head of the caudate nucleus, the preoptic area, the anterior hypothalamus, the ventromedial hypothalamus, the contralateral amygdala, the piriform cortex, the anterior temporal cortex and the insular cortex (Goddard, 1964). Less direct connections with the hypothalamus, subthalamus, mesencephalic tegmentum, hippocampus, and much of the neocortex have also been traced.

The functions of the amygdala have been well surveyed by Goddard (1964), who points out that after bilateral removal of the amygdaloid complex animals frequently remain in a lethargic or depressed condition for a week or more, but then tend to become hyperactive and react extremely readily to environmental stimuli. Food intake may be markedly reduced during the initial phase of depression but may then rise much above normal and lead to obesity. The

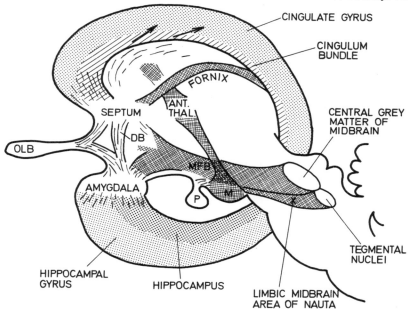

FIGURE 2.7 Schematic diagram of the limbic system and its connections which emphasizes the role played by the medial forebrain bundle (MFB) in linking the limbic lobe and hypothalamus. Other abbreviations: DB, diagonal band of Broca; M, mamillary body; OLB, olfactory bulb; P, pituitary gland. After MacLean (1962).

increased food intake arises from a constant nibbling of food and the consumption of anything that is edible, but the animals do not show a heightened desire for food. Changes in sexual behaviour after destruction of the amygdaloid nuclei have been often described, and hypersexuality in the males of a variety of species reported. This is manifest in an apparent lack of discrimination on the part of the male, in that the animal will attempt to copulate with non-oestrous females, males, members of other species and even inanimate objects such as a toy bear. The few relevant observations in man indicate that the hypersexuality is mild and apparent in an increased holding of hands, exhibitionism, and masturbation. Subsequent lesions placed in the ventromedial nuclei of the hypothalamus, or the septal area, depress the hypersexuality produced by amygdaloid damage in animals.

23

From the psychological point of view the amygdala can be regarded as being concerned with emotionality and the responsiveness to noxious stimuli (Goddard, 1964). Put more directly, this region of the brain may be said to be involved in the genesis of fear. Stimulation of the amygdala in cats leads to defensive or fear reactions, as shown by hissing, vocalization, rage, or flight. Fear responses have also been produced in monkeys and man by amygdaloid stimulation. Bilateral lesions in the amygdala cause a remarkable loss of aggression in savage animals. Wild rats, which previously could not be handled and needed to be approached with circumspection, immediately became placid after destruction of the amygdaloid nuclei and could be handled

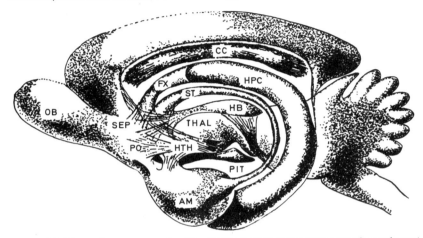

FIGURE 2.8 The major components of the limbic system in the rat as seen in a schematized 'phantom' diagram, projected on a sagittal plane. Most neocortical structures are not indicated, thus revealing the more 'ancient' limbic components of the brain and their connections with the hypothalamus. Cerebellum and lower brain stem on the right. AM, amygdaloid complex; CC, corpus callosum; FX, fornix; HB, habenular complex; HPC, hippocampus complex; HTH, hypothalamus; OB, olfactory bulb; PIT, pituitary gland; PO, preoptic area; SEP, septum complex; ST, stria terminalis; THAL, thalamus. The medial forebrain bundle, olfactory tract, stria medullaris, and habenulo interpeduncular tract are sketched but not labelled. After de Groot (1965).

without gloves immediately upon recovery from the anaesthetic. Placidity has similarly been induced in the cat, lynx, agouti, dog, phalanger, monkey, and man, and has sometimes followed removal of the inferotemporal, piriform and entorhinal cortex without damage to the amygdala. However, these structures supply many afferents to the amygdala. The amygdaloid nuclei are not the only neural structures concerned with emotional responses, for the hypothalamus and septum are also important. Nevertheless, the temporal lobes are highly influential. Occasionally, savage behaviour and rage develops after the placement of lesions in the amygdala and has been attributed to additional involvement of the hippocampus and to irritation of neural tissue at the periphery of the lesion.

In view of the close association between the basal temporal structures and emotion it is not surprising that stimulation of the amygdala elicits many autonomic responses. Changes in heart rate, respiration, blood pressure, the size of the pupil, the blood sugar level, and in adrenal activity have commonly been observed. Possible ways in which emotional stimuli could be fed into the limbic system were considered by Papez (1937) who realized that stimuli of cortical origin could pass to the hippocampus and through the fornix to the mamillary bodies. From there they could travel upwards through the mamillo-thalamic tract to the anterior nucleus of the thalamus and thence back to the cingulate gyrus.

'The central emotive process of cortical origin may then be conceived as being built up in the hippocampal formation and as being transferred to the mamillary body and thence through the anterior thalamic nuclei to the cortex of the gyrus cinguli. The cortex of the cingular gyrus may be looked on as the receptive region for the experiencing of emotion as the result of impulses coming from the hypothalamic region, in the same way as the area striata is considered to be the receptive cortex for photic excitations coming from the retina. Radiation of the emotive process from the gyrus cinguli to other regions in the cerebral cortex would add emotional colouring to psychic processes occurring elsewhere. This circuit would explain how emotion may arise in two ways: as a result of psychic activity and as a consequence of hypothalamic activity' (Papez, 1937).

It will be apparent from this description how readily emotional stimuli can be integrated with neuroendocrine responses generated within the hypothalamus. Although phylogenetically old it seems that evolution of the limbic system has continued in mammals. Thus the number of fibres in the bundle of the fornix running behind the anterior commissure and fed from the hippocampus and gyrus fornicatus increases from 50–60,000 in the rat to 1,200,000 in man. The basolateral cell groups of the amygdaloid complex also undergo a steady increase in volume.

As mentioned earlier, damage to the amygdaloid region in monkeys and cats leads to tameness, docility, and bizarre eating habits, as well as to a form of hypersexuality which is not restricted to the opposite sex or even to the same species. In the light of this information and other considerations MacLean (1962) has suggested that the limbic system is largely concerned with self-preservation and the preservation of the species. He has found that stimulation of various parts of the limbic system, such as the hippocampal projections to the septum, anterior thalamus and hypothalamus, the mamillary bodies, mamillo-thalamic tract, and anterior cingulate gyrus, elicits erection of the penis in the squirrel monkey. Stimulation of the septum or rostral diencephalon which caused penile erection was often followed by after-discharge in the hippocampus which, if repeated, was associated with a change of mood from aggression to placidity.

The Limbic System—Midbrain Circuit

The connections of the hypothalamus with the midbrain which lies immediately behind it are diffuse and no distinct boundary between the two structures can be defined. The lateral hypothalamic region continues past the mamillary bodies into the ventromedial part of the midbrain reticular formation while the more medial hypothalamic zones are continuous with the central grey matter surrounding the aqueduct of Sylvius. When the more anterior connections of the hypothalamus are taken into account it can be seen that this part of the brain forms part of two neural circuits, one of which connects with the limbic forebrain structures and the other with a medial zone of the midbrain. In the eyes of Nauta (1963) the hypothalamus appears as a nodal point in a vast neural mechanism extending from the medial wall of the cerebral hemisphere caudally to the lower boundary of the midbrain so that the functional state of the hypothalamus is continuously influenced by the prevailing activity patterns in the limbic forebrain-midbrain circuit as a whole. The afferent input to this system appears to form part of more diffusely organized conduction pathways ascending from the spinal cord and brain-stem reticular formation which transmits a variety of sensations as well as impulses of visceral origin. It is also argued that since there appears to be no fibre system leading from any part of the limbic system-midbrain circuit directly to somatic or visceral motor neurones it may be that the neural outflow is more particularly concerned with hypothalamic neuroendocrine function.

The Hypophysial Portal Vessels

Knowledge of the existence of the hypophysial portal vessels is of comparatively recent date although they are depicted in the anatomical drawings of Bourgery published in 1845. The first detailed description was given in 1930 by Popa and Fielding and since that time the system has been subjected to painstaking examination. The hypophysial portal circulation is to be found in all vertebrates and is remarkably constant in structure. In the simplest glands of primitive vertebrates the system exists as a mantle plexus which lies between the median eminence on the tuber cinereum of the hypothalamus and the hypophysis so that in general blood passes from the neurohypophysis to the adenohypophysis. Later the neural lobe of the neurohypophysis acquires a partially independent blood supply and three zones of vascularization can be distinguished. These are the hypothalamus, the neural lobe, and the median eminence with the neural stalk and the anterior lobe of the hypophysis, but capillary connections between the regions persist. The portal system in mammals is supplied by branches of the internal carotid arteries which break up into a plexus of capillaries, the primary plexus, which in turn combine to form the portal veins. These pass down the pituitary stalk and break up into sinusoidal vessels within the pars distalis (Figure 2.9). It is because of the similarity between this layout and that present in the liver that the hypophysial vessels are termed 'portal'. However, the liver also possesses a separate, independent, arterial supply which is generally lacking in the pars distalis.

The primary plexus is marked by the presence of numerous loops, whorls, or coils which vary in complexity from species to species. Such loops within the median eminence greatly increase the capillary surface to which the endings of nerve fibres passing into the hypothalamus can be applied and favour the transmission of substances liberated from the nerve endings in the pars distalis. Most prominence has been given to the primary loops in the anterior part of the median eminence (which feed long portal vessels supplying most of the anterior lobe) but it has also been established that the arteries supplying the posterior lobe also give rise to coiled capillaries within neural tissue and that these are drained by so-called 'short' portal vessels which supply an area of the pars distalis lying adjacent to the lower infundibular stem or infundibular process.

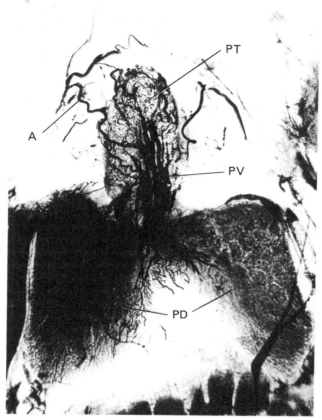

FIGURE 2.9 Photomicrograph of a horizontal section through the median eminence and pituitary gland of a rat, following perfusion of the blood vessels with indian ink. Small arterial twigs (A) supply the pars tuberalis (PT) and its associated vascular plexus, together with the loops of the primary plexus in the median eminence. Portal vessels (PV) run from the pars tuberalis and median eminence along the ventral surface of the infundibular stem to the pars distalis (PD). From Green and Harris (1947).

All the blood reaching the anterior pituitary gland in mammals (with the only known exception being the rabbit) travels down the pituitary stalk. Both groups of portal vessels, the long and the short, are fed by capillary beds in neural tissue but they appear to supply separate zones in the pars distalis. It is possible that the portal vessels are so arranged that substances released from the nerve endings can be delivered to specific territories of glandular tissue but evidence on this point is difficult to collect. Nevertheless, it is known that the cells secreting individual pituitary hormones tend to be grouped in clusters in certain areas which could be supplied by specific portal vessels. Hemisection of the pituitary stalk leads to unilateral necrosis of the gland but this finding may simply indicate that there are no anastomoses between the portal veins.

Since neurohumoral agents are believed to be transferred to the portal vessel system within the median eminence considerable effort has been put into studies of the anatomy of this region in the hope of visualizing the nerve terminals together with their stores of releasing substances. Excellent work has been done with the aid of the electron microscope and nerve terminals have indeed been shown to be closely applied to the capillaries of the primary plexus. The terminals contain vesicles of various kinds which could be liberated to gain access to the portal vessel blood through tiny pores in the endothelium of the capillary wall. The nature of the several types of inclusions has not been determined, although there is evidently a morphological basis for the observations shortly to be described (Rinne and Arstila, 1965/66).

Many pharmacologically active substances are present in the hypothalamus, among which the catecholamines are prominent. Noradrenaline, dopamine, and their oxidative enzymes dopamine oxidase and monoamine oxidase are present in relatively high concentration. Dopamine is present in large amount in the dorsomedial nuclei and in the external layer of the median eminence, whereas noradrenaline is more widely distributed in the hypothalamus, as is cholinesterase, the indicator of the presence of cholinergic cells. Many monoaminergic and cholinergic pathways can be traced through the hypothalamus, but while the presence of acetylcholine in brain tissues is to be expected, little is known of the function of monoaminergic neurones in the brain (Shute and Lewis, 1966).

Quite recently, a highly sensitive fluorescence method for localizing monoamines in the brain has been developed in Sweden (Falck, 1962; Falck and Owman, 1965), and applied to the hypothalamus. In the mouse, for example, fibres with a varicose appearance are restricted to the part of the median eminence covered by the primary plexus, and the capillaries are partly or completely surrounded by structures containing a high concentration of monoamines. The first traces of monoamines become visible in the presumptive median eminence of foetuses during the last 3 days of gestation, but typical monoamine containing fibre endings are not found until the end of the first post-natal week, with the adult condition being reached two weeks later. The supraoptic, paraventricular, and dorsomedial nuclei of the hypothalamus possess a prominent plexus of fluorescing fibres, while the arcuate nuclei, among others, contain fluorescing cell bodies; the ventromedial nuclei are almost devoid of monoaminergic fibres. Despite the fluorescence of the supraoptic and paraventricular nuclei, the fluorescing material is not associated with the typical neurosecretion described in chapter 5 (Bjorklund, Enemar, and Falck, 1968). Gonadectomy in the rat increases the catecholamine content of the hypothalamus, and electron microscopic studies of the arcuate nuclei have shown that after this operation there is a general hypertrophy of the arcuate neurones together with an increase in the number of granulated vesicles in the axons and in the outer zone of the median eminence. The granulated vesicles are believed to contain catecholamines (Zambrano and de Robertis, 1968a). The catecholamine content of nerve cells in the tuberal region of the hypothalamus of rats (mostly in the anterior three-quarters of the arcuate nuclei) has been reported to vary in phase with the oestrous cycle from day 1 of dioestrus to the day of oestrus, when there was a marked fall. Only slight changes in the intensity of fluorescence, and a different rhythm, were observed in a control area of cells in the substantia nigra (Lichtensteiger, 1969). Assay of the noradrenaline content of various regions of the hypothalamus has indicated that the concentration in the anterior region showed most change during the oestrous cycle of the rat. It was highest at prooestrus, decreased after ovulation and was lowest at oestrus. The noradrenaline content of the anterior hypothalamus rose after gonadectomy, and when tritiated tyrosine was given to assess the rate of biosynthesis of noradrenaline in the hypothalamus, turnover of this catecholamine was found to be increased after spaying (Stefano and Donoso, 1967). Somewhat different results have been obtained with measurements of the overall noradrenaline content of the hypothalamus. The concentration was lowest at prooestrus, had risen on the day of oestrus and was maximal on the second day of dioestrus. The noradrenaline content fell gradually after spaying but treatment with oestradiol benzoate restored the level to that recorded at dioestrus. Parallel observations on monoamine oxidase activity in the hypothalamus showed that this was greatest in the supraoptic and paraventricular nuclei at prooestrus, whereas in the dorsomedial nuclei, ventromedial nuclei, and infundibular region peak activity occurred at oestrus (Kurachi, Iwata, and Hirota, 1968). Other workers (Kobayashi, Kobayashi, Kato, and Minaguchi, 1966) found that the choline acetylase activity of the anterior hypothalamus fluctuated during the oestrous cycle while that of the posterior hypothalamus remained constant. Enzyme activity was highest at metoestrus and declined through dioestrus and prooestrus to be lowest at oestrus. The monoamine oxidase activity of the hypothalamus also varied, but this was highest at prooestrus and fell as the cycle progressed, to be minimal at dioestrus. In this case the activity of the anterior hypothalamus remained constant while that of the posterior hypothalamus fluctuated. Ovariectomy increased the choline acetylase and monoamine oxidase activity within the hypothalamus and these changes could be suppressed by treatment with oestradiol benzoate. Progesterone elicited no definite change in the enzyme activity of the hypothalamus.

At the present time, many studies of the hypothalamic content of catecholamines, and of other substances, in a variety of circumstances are under way and only representative examples have been mentioned to illustrate the kind of information being collected (Fuxe and Hökfelt, 1969). However, in the absence of facts concerning the rate of synthesis and of turnover of the material in question it is difficult to relate the changes observed to the secretion of active compound, for the release can only be guessed at. The need for the additional insight readily becomes apparent when it is realized that if the rate of release is matched by increased synthesis then there will be no change in the amount of material stored, whereas if the rate of synthesis exceeds that of release the stock will increase despite the enhanced liberation of active compound. Only if there is a time lag between the release of stored material and the formation of new will discharge become apparent. These considerations are, of course, of general application and will be familiar to those, for instance, who have tried to assess the rate of secretion of pituitary hormones by the assay of the stored factor in the hypophysis. Such changes as are observed in the hypothalamic content of a particular substance may not directly reflect changes in the activity of the nervous system, but may arise secondarily through the feedback action of target organ hormones. The monoamine oxidase activity in the hypothalamus, mentioned above, is influenced in this way by oestrogen.

The noradrenaline content of the brain can be depressed in several different ways, including inhibition of the synthesis of the amine (as with α-methyl-tyrosine) or interference with the storage of the transmitter (as with reserpine). The drug iproniazid prevents the depletion of brain noradrenaline by inhibiting the enzyme which oxidatively de-aminates noradrenaline and adrenaline, so that pretreatment with this compound blocks the pseudopregnancy elicited by reserpine. These results have been taken to indicate that an adrenergic tonus originating in the hypothalamus influences gonadotrophin secretion through an effect on the specific releasing and inhibitory factors. It has been suggested that when sympathetic activity is stimulated the hypothalamic production and release of FSH-RF, LH-RF, and PIF are enhanced, whereas the converse situation, favouring prolactin release, is produced when the sympathetic tone is reduced or absent, as after treatment with a noradrenaline depleting drug (Coppola, 1968). There are indications that it is the relative availability of the amines that is important from the physiological point of view, rather than the absolute concentration, but the whole concept, though attractive, awaits proof.

3

Methods in Neuroendocrine Research

The majority of the techniques employed in neuroendocrine investigations stem from basic neurophysiological or endocrine research, and will be familiar to students of those subjects. Nevertheless, it is useful to discuss the advantages and disadvantages attached to the employment of a particular technique, for in this way it is much easier to follow the development of the field and to assess the validity of the results obtained.

Hypophysectomy

After the demonstration by Philip Smith in 1927 of the feasibility of the para-pharyngeal approach to the pituitary gland in the rat with subsequent aspiration of the organ, this technique for hypophysectomy was adopted by innumerable workers, for in the rat the procedure is relatively simple. With practice such operations can be performed in two or three minutes by a technician and assistant. However, extreme care must be exercised because fragments of the pituitary gland are readily left behind in the sella turcica with consequent ruin of an experiment. Fragments of pituitary tissue may adhere to the walls of the sella turcica or may be out of sight in those species (dog, monkey, and man) where an oral or subtemporal approach has been adopted. The completeness of hypophysectomy is best checked by examination of serial histological sections through the sella turcica and adjacent tissues, although some have relied upon macroscopic examination of the operation site or study of the arrest of growth and atrophy of the target organs that characteristically follows this operation. Both latter methods are unreliable. It should also be borne in mind that it is not possible to remove the glandular tissue of the pars tuberalis without damaging the median eminence of the tuber cinereum.

Pituitary Stalk Section

Once the effects of hypophysectomy seemed reasonably clear the next logical exploratory step was to study the consequences of division of the pituitary stalk. Some experiments involving pituitary stalk section were carried out in advance of sound knowledge of the changes wrought by hypophysectomy but the proper interpretation of this pioneer work has had to wait upon later developments. As Jacobsohn (1966) has truly remarked, 'for those who during the last thirty years have been engaged in studies of the hypothalamic control of the pituitary gland, the term 'pituitary stalk section' has connotations that can hardly be conveyed to others. From the intention to denervate the gland a story arose full of contradictions and hard work in many minds and skilful hands were required until the puzzling role of the hypophysial portal vessels accompanying the hypophysial stalk was clarified.' Much of the confusion arose because of the supposition that the depression of pituitary function following stalk section was due to interruption of nerve fibres passing along the stalk from the hypothalamus to the attached gland. The functional importance of the portal vessels and the ease with which they regenerate after damage had not been grasped, so that only rarely was a barrier inserted between the brain and the pituitary gland. This is now an inescapable requirement in all procedures designed to permanently interrupt the hypophysial portal vessels.

The pituitary stalk can be approached parapharyngeally, transpalatally, transorbitally, or subtemporally. The parapharyngeal and transpalatal approaches are less difficult than the subtemporal and avoid damage to the brain but have the disadvantage that barriers inserted beneath the hypothalamus can be extruded and that some disorganization of the tissues in the region of the pituitary stalk can ensue. On the other hand the subtemporal route involves major surgery such as the removal of part of the skull and manipulation of the brain, together with delicate operations within a small field of view, but has the advantage that a barrier located between the hypothalamus and pituitary gland is not readily displaced and the anatomy of the area does not become distorted. Since the transorbital path to the hypothalamus involves removal of an eye this technique is seldom used. Just as for hypophysectomy it is essential to control the success of any operation by serial section of the hypothalamo-pituitary region, preferably after perfusion of the vascular system with indian ink.

Transplantation of the Pituitary Gland

Permanent interruption of the hypophysial portal vessels can be readily achieved by transplantation of the pituitary gland to a site far from the sella turcica. The simplest procedure is to implant pituitary tissue from a donor animal into an hypophysectomized host, but unless the host and donor are very closely related antibody formation can occur and a homograft reaction develop with consequent rejection of the implant. However, certain locations such as the anterior chamber of the eye and the brain, seem to be immunologically privileged in that homografts placed in these sites are tolerated and may function for extended periods of time. This phenomenon has proved of value in the comparison of the activity of pituitary grafts placed in contact with the median eminence, under the temporal lobe of the brain or inserted directly into the hypothalamus (see chapter 10). The development of an antibody reaction can be avoided by autotransplantation of pituitary tissue, although in this case the gland suffers from the manipulations involved in the transfer to a new position. Loss of pituitary tissue inevitably occurs and it is fortunate that normal endocrine function can be supported by less than 50% of the normal volume of the gland. Indeed, it has proved possible to compare the function of anterior pituitary tissue autotransplanted to the kidney in rats with that seen after replacement of the graft beneath the median eminence. Despite the double insult oestrous cycles were resumed, after a connection between the hypothalamus and anterior pituitary gland was restored, so indicating that the activities of the grafts approximated to normal (Nikitovitch-Winer and Everett, 1958). Direct evidence of the action of extracts of median eminence tissue upon the pituitary gland has been provided by Evans and Nikitovitch-Winer (1969). Autotransplants of pituitary glands of female rats were made to the capsule of a kidney and treated with acid extracts of median eminence or of cerebral cortex by prolonged infusion of the solution through a catheter chronically situated in the renal artery. Infusion could be maintained for many days and it was found that gonadotrophin secretion occurred when median eminence extracts were employed but not when extracts of cerebral cortex were applied. Cytological changes in the pituitary grafts and follicular development in the ovaries were observed, and there was signs of oestrogen secretion, but ovulation never took place. Perhaps surprisingly, there was little evidence of an enhanced secretion of adrenocorticotrophic or of thyrotrophic hormone but this could have been due to inactivation of the appropriate releasing factor during the preparation of the extracts.

Autografted pituitary tissue may support a minimal level of thyroid function, but this is much less than that achieved by neonatal glands (Purves, Sirett, and Griesbach, 1965/66). Multiple grafts of pituitary tissue made into young rats function sufficiently to maintain the weights of target organs above those of completely hypophysectomized controls and to support body growth, as distinguished from the mere maintenance or increase in body weight by the accumulation of fat. This effect is not due to the release of stored hormone from the implanted pituitary glands because destruction of the median eminence in such animals inhibited growth and reduced the weights of the testes (Beddow and McCann, 1969). The more likely explanation is that sufficient amounts of hypothalamic releasing factors can be liberated into the circulation of hypophysectomized animals to affect pituitary tissue located far from the sella turcica. The considerations outlined above apply only to the pars distalis and to target organs free from direct neural control. The pars neuralis of the hypophysis and the adrenal medulla do not take kindly to transplantation.

Brain Lesions

In order to destroy, stimulate, or record from a particular part of the brain, that area must be located with accuracy. Superficial areas, such as the cerebral or cerebellar cortex, can be exposed under visual control but guidance is needed when deeply lying structures are sought. This is usually provided by a stereotaxic machine, which rigidly holds the head in a predetermined position and, with the aid of a suitable atlas of the brain, enables electrodes or other devices to be introduced along the appropriate coordinates to a chosen depth. In the rat, cat, and man the reference plane from which all measurements are taken is the line passing through the two auditory meatuses. This provides the horizontal reference parameter, with antero-posterior measurements being related to a vertical plane drawn at right angles to the reference axis. It is necessary to fix the nose of the animal in a standard way, for the head can rotate about the ear bar axis, but the position adopted varies from atlas to atlas. In some the head is tilted upwards, above the horizontal, in others the nose is pointed downwards. By varying the position of the nose access to some parts of the brain stem is facilitated, while other structures, important for the work in hand, can be avoided. Because the auditory canals in the rabbit are not horizontal and cannot be used for fixation of the head, the skull is usually gripped by clamps over the posterior ends of the zygomatic arches. In this case the position of structures within the brain can be related to the point on the surface of the skull (bregma) where four sutures join. Further information about stereotaxic techniques and a bibliography of stereotaxic atlases has been provided by Rowland (1966) and de Groot (1966b).

Destruction of a limited volume of brain tissue can be brought about in a variety of ways. Chemicals causing necrosis of neural tissue can be applied or injected, or specially shaped knives or tubes can be inserted into the brain. Where the tissue to be removed is at or near the surface of the brain, suction can be applied for ablative purposes. More elaborate procedures include the implantation of radon seeds or the application of focussed ultrasonic waves, but the most widely employed technique of all is simple electrolysis.

For the production of electrolytic lesions a direct current is passed between the bared tip of two insulated wires thrust into the brain, when the tissue lying between them suffers damage, or between a single fine brain electrode and an 'indifferent' large electrode inserted into the mouth, rectum, or in contact with a large area of skin. In the latter case lesions of varying size can be produced depending upon the size of the current and the length of time the current is applied, but the occurrence of several destructive processes occurring simultaneously can cause complications. This is because the passage of current causes injury by electrolysis, by coagulation through the heat developed, and by the pressure effects of the gas evolved at the electrode tips. If an alternating current of radiofrequency is used, purely thermal lesions can be made and these are often preferable. The electrolysis accompanying the passage of a direct current can cause unexpected effects. When steel electrodes are employed iron is carried into the neighbouring zone and can stimulate brain tissue. Short-lived mating behaviour has been observed after the placement of such lesions in the preoptic area and is probably due to this effect while other lesions in the preoptic area have caused ovulation in rats. Similar lesions produced by radiofrequency, thermocautery, or by electrolysis using platinum electrodes did not induce ovulation. In view of these considerations it would be very useful to reversibly inactivate the neurones in small areas of the brain in order to check whether the changes seen with inactivation wear off with removal of the constraint. One possible method is by local cooling of the brain, but this is hardly practicable for long continued, chronic, experiments.

The effects of an experimental lesion in the brain do not necessarily correspond to those accompanying pathological involvement of the same area. A tumour of the hypothalamus, which may grow quite slowly, can be quite extensive without producing signs of hypothalamic dysfunction. But surgical removal of the growth may be followed by marked hypothalamic derangement. In goats it has been found that radiofrequency lesions placed in the preoptic area interfere with the processes of heat regulation, for strong shivering and intense vasoconstriction appeared at once and the body temperature rose. Death could ensue within hours. On the other hand, when destruction of the preoptic area was accomplished slowly by irradiation with a beam of protons, hyperthermia was less extreme and the animals survived until they were deliberately sacrificed many days later (Andersson, Gale, Hökfelt, and Larsson, 1965).

The changes in brain or endocrine function seen after destruction of a cell group or of a well-delimited area do not necessarily indicate that the region damaged is involved in the control of that function. Unsuspected groups of nerve fibres passing through the zone may have been interrupted, or the blood supply to another nucleus or cell group may have been damaged inadvertently. The possibility of irritation of the neurones at the periphery of the lesion must also be taken into consideration.

Since the brain is made up of two basically equal halves with pairing of almost all components it follows that lesions placed in the brain should be paired and symmetrical. Otherwise the nucleus remaining undamaged may well be able to compensate for the loss of its partner. Unfortunately precision in this regard is difficult to obtain and it is perhaps surprising that so much useful information has accumulated. In the same vein it may be pointed out that lesions are seldom confined to neurones of a single type or to a group of cells subserving a single function. The cells of the supraoptic and paraventricular nuclei in the hypothalamus do differ significantly from their neighbours and these nuclei can be singled out for destruction. When this is done diabetes insipidus ensues and it is clear that these nuclei are concerned with the control of neurohypophysial function. Additionally, efforts have been made to relate the paraventricular nuclei to the regulation of gonadal function, for lesions in the neighbourhood of these nuclei (where well defined nuclei are lacking) have long been known to cause constant vaginal oestrus in rats and the disappearance of corpora lutea from the ovaries. Since small symmetrical lesions placed caudally to the paraventricular nuclei also induced the same changes it seemed that nerve fibres running from the nuclei to the median eminence could have been destroyed but when lesions were restricted to the paraventricular nuclei themselvés no gonadal changes ensued, so that the paraventricular nuclei do not seem to be important in the control of gonadal function (Olivecrona,

1957). This work illustrates the difficulties that arise in attempts to distinguish between the effects of destruction of nerve cells and those following interruption of fibres passing through the region. The results of investigations aimed at the mapping of the hypothalamus in terms of the areas concerned with the secretion of the separate pituitary trophic hormones have established that much overlap occurs. Thus the release of adrenocorticotrophic hormone is controlled through a relatively diffuse area which involves the anterior and middle portions of the median eminence in the dog and most of this structure in the rat. This field overlaps with the area governing thyrotrophic hormone release located above and behind the optic chiasma and with the zone known to be concerned with gonadotrophin secretion. When it is realized that despite this overlap in representation the output of one pituitary hormone can be varied quite independently of another, and that when the release of adrenocorticotrophic hormone is enhanced that of thyrotrophic hormone is often depressed, it is less surprising that the interpretation of the effects of lesions is no easy task.

It might be anticipated that depression of brain or endocrine function always follows damage to some part of the nervous system. This is not so, for the changes, or lack of change, observed result not only from the loss of the tissues destroyed but also from the continued activity of the undamaged parts. In some cases nervous or endocrine function is increased through the removal of an inhibition exercised by the region involved in the lesion. Two kinds of inhibition can be visualized: that produced by the suppressive feedback action of target organ hormones on the brain, and that which is fundamentally neurogenic in origin. The persistent vaginal oestrus seen in rats and guinea-pigs following the placement of appropriately located hypothalamic lesions has been attributed to an over-secretion of follicle-stimulating hormone due to reduction in the intensity of the feedback action of oestrogen. Because of loss of sensitive hypothalamic tissue the intensity of the inhibition exerted by the gonadal hormone may be thought to be lessened. There is evidence, too, that lesions placed in the posterior median eminence and hypothalamus, or the implantation of oestradiol, both increase the secretion of prolactin, probably by removing an inhibitory influence exerted through that region. Lesions in the posterior hypothalamus are reported to favour the release of increased amounts of adrenocorticotrophic hormone under stressful conditions (chapter 7).

Electrical Stimulation of the Brain

In theory, electrical stimulation of a part of the brain is relatively easily achieved. All that is necessary are two wires, insulated except at the tips, which are inserted into the desired area and through which the stimulus is applied. In practice, many refinements are needed and a number of reservations need to be borne in mind in the evaluation of the results. Since the various parts of the brain are not equally excitable changes in the nature of the stimulus can alter the response observed from stimulation of a particular area. Thus it becomes important to select the parameters appropriate for the excitation of the structure to be examined. The parameters include the waveform of the stimulus (whether it is a sine wave, a rectangular pulse, a unidirectional or a bidirectional pulse, or a spike), the duration of the pulse, the frequency of repetition and the duration of the train of pulses, not to speak of the voltage and current to be applied. In one series of experiments involving stimulation of a single locus in the amygdala, behavioural changes previously attributed to activation of different nuclei could be produced merely by altering the characteristics of the stimulus. Care must also be taken to avoid the inadvertent production of a lesion at the site of stimulation. If electrical stimulation is carried out in anaesthetized animals then the nature and depth of anaesthesia can influence the results. Barbiturate anaesthetics depress neuroendocrine mechanisms to a far greater degree than does ether.

Electrical stimulation of the brain is not necessarily highly selective in character. Once the current is switched, on a wide variety of cells, which, if physiologically interconnected, normally fire in sequence now fire simultaneously so that there can be a concurrent excitation of inhibitory and excitatory neurones. This phenomenon is well known in studies involving stimulation of the motor areas of the cerebral cortex where some time elapses between the stimulus and the motor response due to interaction of excitatory and inhibitory pathways which are simultaneously activated. Movement occurs only when one facet of the response achieves dominance. And not all the cells stimulated are necessarily concerned with the function under study. Electrical stimulation of the anterior hypothalamus will evoke the release of several pituitary hormones at once, although under physiological circumstances they are seldom released together. Autonomic mechanisms represented within the hypothalamus are also activated.

It is often difficult to decide upon the volume of brain to be excited. For precision it is preferable to stimulate only a tiny region so that spread to neighbouring and irrelevant structures can be avoided. On the other hand it is essential to stimulate enough of a physiologically important zone to bring about some change in bodily function. Unless a certain threshold level in response is surpassed erroneous negative results will be recorded. In practice it has proved to be extremely difficult to excite all the cells in a zone concerned with a particular function and the useful observations that have been made are perhaps remarkable for their number.

As will be apparent in subsequent chapters, it is often difficult to distinguish between stimulation of cell bodies or of fibres passing through a particular region. The parameters of stimulation used in evoking a response are of value in deciding questions of this kind but the answer obtained is not necessarily fully convincing because the response of nervous tissue to stimulation can change because of the movement of ions across nerve membranes due to the electrical field set up, and because of the effects of chemical products of electrolysis upon the nerve membranes. There is also the heating effect of the passage of the current to be taken into consideration as well as local changes in blood flow that may indirectly influence neuronal activity.

The electrodes employed are subject to certain limitations. The materials used in their construction should not be harmful when applied to brain tissue and the passage of current through the tips should not liberate toxic substances within the brain. Minimal damage to the brain should be caused during insertion of the electrodes and they should not be so large as to cause displacement of brain tissue in that neighbourhood so that pressure effects ensue. On the other hand, if the electrodes are very thin and delicate and are removed before the subject is killed localization of the position of the electrode tips can prove difficult. Where possible it is preferable to fix the brain with the electrodes left in place so that a well marked track remains when the electrodes are subsequently removed. Alternatively, with steel electrodes the location of the electrode tips can be marked by the passage of a small direct current with the electrode under study being made positive. Ferric ions then pass into the tissue and the Prussian blue reaction developed by immersion of the tissue in a solution of potassium ferrocyanide. If the electrode tip is large enough the passage of a greater direct current will produce a small lesion which can be traced histologically. For minimal electrochemical stimulation of the brain when concentric steel electrodes are used Terasawa and Sawyer (1969) indicate that the stimulating tip should be the cathode and that the waveform should be monophasic with individual pulses lasting less than 0·5 ms and having a current intensity below 0·3 mA. Under the conditions of their experiments stimulation of the medial preoptic region of the rat brain with 0·5 ms monophasic waves at 200 c/s, applied for 15 seconds on and 15 seconds off for 30 minutes at 0·1 mA caused ovulation without the deposition of iron.

Perhaps the greatest degree of refinement in experiments involving electrical excitation of the brain is achieved when the individual under test is able to switch the stimulus on at will. Work on rats was pioneered by Olds (1956) who inserted electrodes into the brain and coupled them through a stimulator to a lever in a Skinner box so that every time the lever or treadle was pressed a tiny region of the brain was excited. When electrodes were implanted into the classical sensory or motor systems the response rates stayed at the chance level of about 10–25 presses an hour, but when the electrodes were in the hypothalamus the animals would stimulate themselves from 500 to 5000 times an hour and even ignore food to carry on stimulation for 24 consecutive hours. There is little doubt that the rat finds such electrical stimulation infinitely rewarding, but there is no way of knowing what sensation is being experienced. This work provoked many other investigations so that there is now a large literature describing the changes in behaviour and physiological function that occur with self-stimulation, as well as the effects of treatment with drugs and hormones on the basic reaction (Grossman, 1967a). However, although most commonly effective self-stimulation is obtained with electrodes in the medial forebrain bundle and lateral hypothalamus the anatomical basis of the response still awaits proper analysis (Wetzel, 1968).

Chemical Stimulation of the Brain

Some of the problems which arise with electrical stimulation of the brain can be avoided by employing drugs, for it might be expected that more specific responses can emerge and that the whole of an active area can be influenced when a chemical agent is given systemically. But, in turn, problems of the specificity of action of drugs emerge which render interpretation of the results difficult. Ovulation in the rat or rabbit can be blocked by the appropriately timed administration of a variety of neurally active drugs of different types. Since these include anti-adrenergic, anti-cholinergic, anti-histaminergic, tranquillizing, and anaesthetic agents, the conclusion to be drawn is simply that the systems governing ovulation are complex and include several pharmacological steps.

The pharmacological approach to the study of neuroendocrine function can be refined by applying the chemical directly to the area of interest. This can be done by local injection of a solution of the material, by the extrusion of a tiny pellet of substance from a hollow needle inserted into the brain, or by the implantation of needles within the brain which are coated with the substance. Control experiments are of extreme importance in work of this kind for damage to the brain can be readily induced. Some indication of the volume of brain tissue affected by the drug can be gained from control experiments in which the drug is mixed with a dye and the fate of the dye followed. However, it is always difficult to find a suitable dye which has similar physico-chemical properties (solubility, diffusion rate, capacity to cross cell membranes) as the experimental compound. Chemical stimulation of the brain with a particular substance can produce different responses from species to species. Nevertheless, positive results can be quite informative, such as the finding that the injection of noradrenaline into the brain just above the hypothalamus causes eating, whereas the application of acetylcholine to the same site leads to drinking. The specificity of the results obtained with this technique may be regarded as remarkable when it is realized that the entire bilateral volume of the rat hypothalamus is about 8 microlitres and that when a unilateral injection of 1 microlitre is made one-fourth of the ipsilateral side must be displaced to accommodate a volume equal to a quarter of the structure (Myers and Sharpe, 1968). Pressure effects can be avoided by injecting solutions directly into the ventricles of the brain, but the responses observed when this route is used may differ from those elicited by injections made directly into the brain substance. Carbachol, for example, causes drinking when applied to the hypothalamus but not when injected into the ventricular system. Adrenergic drugs, on the other hand, evoke eating when administered in either way. In some experiments tubular electrodes have been implanted into the brain so that the effects of electrical or chemical stimulation of the same locus could be compared (Baxter, 1967). The results differed according to the kind of stimulation applied, in that the behavioural changes produced by electrical stimulation varied with the location of the electrode, whereas that following chemical (carbachol) stimulation did not alter. When injected into amygdaloid or hippocampal sites from which no emotional behaviour was elicited by electrical stimulation, carbachol produced behaviour similar to that obtained from the hypothalamus. These results can be explained by excitation of different neural systems, but diffusion of the drug away from the site of injection may be of more importance (Baxter, 1968).

Localized treatment of regions of the brain with hormones has proved enlightening in recent years, with attention being focussed particularly upon the hypothalamus. The source of hormone has varied, from the endocrine tissue itself to the pure material. Fragments of an ovary, or of a thyroid gland have been implanted into the hypothalamus to locally raise the concentration of ovarian or thyroid hormone. While this technique has the advantage that the effect of endogenous hormones is followed, pressure effects produced by the graft as it grows are difficult to avoid. When hormones are applied in other forms, whether as solids, liquids, or gels, the fact remains that from the point of view of the neighbouring neural tissue the concentration of hormone released from the graft is unphysiologically high. A recent improvement in technique now makes it possible to treat parts of the brain with hormones for short periods by inserting tubes loaded with the hormone through one or more guide barrels permanently located on the skull. Once the guide barrel has been fitted to the calvarium further anaesthesia of the animal during the later introduction of the hormone-tipped tubes is unnecessary. However, paradoxical responses to the local application of hormones can emerge.

Thus, testosterone injected into the medial preoptic area of the male rat induces maternal behaviour as indicated by retrieval activity. The same hormone applied to the lateral preoptic area evokes male sex behaviour, even in female rats.

Some insight into the functional activities of the hypothalamus has also been gained from measurement of the content of stored materials. Investigations of this kind have proved both rewarding and frustrating: rewarding in that it has emerged that the hypothalamus contains high concentrations of pharmacologically active substances, which include acetylcholine, adrenaline, 5-hydroxytryptamine, and histamine, but frustrating in that the significance of these high concentrations remain obscure. As an elaboration of this procedure parts of the hypothalamus have been cultivated *in vitro* and the secretory capacities of the tissue followed. The uptake of hormones, as well as of other substances, can be followed by radioactive tracer techniques. The injection of labelled oestrogen into spayed female cats causes behavioural oestrus and some indication of the parts of the hypothalamus involved in this response can be derived from the location of the tracer with the aid of autoradiographs. However, it has also become evident that the radioactivity initially present in the brain has disappeared by the time the cat responds behaviourally. Changes in the size of the nuclei of the cells in the hypothalamus have been found to follow interference with a variety of endocrine functions. Unfortunately, interpretation of these results is complicated because some nuclear groups respond to almost all types of endocrine disturbance.

Stimulation of peripheral afferent end organs, such as mechanical manipulation of the vagina or cervix, or receptors in the skin can alter the electrical activity of the hypothalamus and the response is influenced by hormones. Hypothalamic activity can be traced by recording the resting oscillations of the EEG type, by examining the slow d.c. changes, by picking up the potential changes evoked by electrical stimulation of an afferent pathway, or by following the action potentials of single cells or units (Beyer and Sawyer, 1969). The action potentials of hypothalamic neurones tend to be smaller than those recorded in the cortex, thalamus, or hippocampus, and the majority of neurones show a more regular pattern of firing than that observed with cortical or thalamic cells. Discharge rates of 1–10 per second are most common, but neurones which fire as rarely as once in 10 minutes and others which discharge at rates up to 40 per second have been found. Great difficulty arises in attempts to interpret the results when changes in unit activity within the hypothalamus are followed. Considerable variations of 'spontaneous' activity and responsiveness can be observed in widely separated cells. Such lability in unit discharge argues against the concept of separate functional centres existing within the hypothalamus. When the response to mechanical probing of the cervix in rats was followed, 60% of the hypothalamic units showed an alteration in firing rate, with the great majority discharging more rapidly. However, the vast majority of cells also reacted to the other test stimuli, which were pain, cold applied to the tail, or olfactory cues. Sensory interaction was observed in that an olfactory stimulus was able to block the response to cervical probing, while cervical probing could block the response to pain or cold. The responsiveness of the neurones altered with the stage of the oestrous cycle and oestrous rats yielded more than twice the percentage of neurones showing inhibitory responses to pain, cold, or cervical stimuli than did dioestrous rats (Cross, 1964; Cross and Silver, 1966). Quite recently it has become possible to record the activity of single neurones in the hypothalamus, preoptic area, hippocampus, and elsewhere in unanaesthetized, unrestrained animals (Komisaruk and Olds, 1968).

Interpretation of the activity of hormones on single units in the hypothalamus can be complicated because of the possibility that the firing rate may change, not directly through the action of a hormone, but secondarily because of a change in input due to hormonal action elsewhere. This difficulty can be overcome by applying the hormone, or other chemical agent, very close to a neurone by microelectrophoretic techniques (Steiner, Ruf, and Akert, 1969). In this way, using a multibarrelled micropipette, minute amounts of soluble steroids can be introduced into the immediate extracellular environment of single cells and the local response recorded. The responding neurone can then be marked with a vital dye without moving the micropipette and later located in histological sections of the brain. Thus the position of steroid-sensitive cells in the brain can be plotted and the nature of their response to hormones determined.

Changes in the activity of parts of the brain can be brought about by other means. For example, the temperature of part of the hypothalamus may be raised or lowered with the aid of a thermode (a sealed tube perfused with liquid held at the desired temperature). This technique has proved its value in studies of the control of body temperature and its relationship to the control of thyroid function. The existence of osmoreceptors within the hypothalamus is inferred from the fact that the injection of hypertonic saline into the region of the supraoptic nuclei increases the secretion of antidiuretic hormone from the neurohypophysis; local injection of hypotonic saline reduces the output of this hormone. Discrete amounts of various anions, cations, or zwitterions can be introduced into the brain by means of a bead of ion-exchange resin loaded with the appropriate agent. In this way diffusion is limited and the bead can be withdrawn later to determine whether any changes observed are reversible (Butcher and Fox, 1968).

Isolation of the Hypothalamus

Experiments involving the placement of lesions in, or stimulation of, the hypothalamus suffer from the disadvantage, mentioned earlier, that it is difficult to determine whether the effects observed are due to disturbance of the connections linking other parts of the brain with the hypothalamus, or to direct disruption of some hypothalamic process. For this reason study of the functional capacity of islands of hypothalamic tissue deprived of connections with other neural structures is currently attracting increasing attention (Halász, 1969). Originally, isolation of the hypothalamus was achieved by removing much of the rest of the brain but the care of such brainless animals posed many problems. Nevertheless, it was found that in the total absence of the hypothalamus the ovaries of both the cat and the rat became atrophic, whereas removal of the cerebral cortex, basal ganglia, and dorsal thalamus, followed by transection of the brain stem between the mamillary bodies and pons was compatible with gonadotrophin secretion, so that the ovaries of animals surviving for long periods (90–360 days) were of normal weight, and histological appearance (Woods, 1962). Attempts have been made to isolate the hypothalamus in other ways involving major surgery or the use of a proton beam produced in a cyclotron (de Groot, 1962) but the major advance came when Halász and Pupp (1965) devised a small knife which could be introduced into the brain under stereotaxic guidance and manipulated so as to cut all the connections around the hypothalamus. Little damage is done to the hypothalamus or neighbouring structures so that the survival of the experimental animals is facilitated. The precise changes in hypothalamic function observed depend upon the precise limits of the island. Animals in which large islands incorporating most of the hypothalamus have been produced, do less well than others with smaller islands bounding the median eminence because the mechanisms controlling body temperature and food intake become deranged. When these complications are avoided, it can be seen that gonadotrophin secretion continues, though not necessarily cyclically in females, while that of adreno-corticotrophin may be enhanced, probably through removal of inhibitory influences exercised through the hypothalamus. A number of the consequences of hypothalamic isolation can be reproduced by making a cut restricted to the anterior border of the hypothalamus. The precise position of the cut has proved to be important, so indicating that the effects do not stem solely from the interruption of afferent fibres. For if that were the case division of the fibres almost anywhere along their length should bring about the same response. The unit activity of freshly prepared large islands of hypothalamus is currently under investigation, because the firing rate of cells in the decerebrate preparations made for short-term experiments can be followed in the absence of anaesthesia and the immediate effects of hormone administration traced (Cross and Kitay, 1967). This is of particular interest because any changes observed must be due to a direct action of the hormone upon the hypothalamus.

4

The Adrenal Medulla

Unlike the cortex of the adrenal gland, the medulla is not essential to life. This can be shown very clearly in animals in which the adrenal medulla and most of the cortex has been removed, for then the remaining cells of the cortex multiply to restore normal adrenocortical function and there are no sustained ill effects under reasonably equable conditions. On the other hand, the output of hormones by the adrenal medulla is greatly increased under conditions of stress, as is the output of hormones by the adrenal cortex, and this response has been taken to indicate that the medulla serves to reinforce the response of the autonomic nervous system to noxious stimuli. After loss of the medulla of the adrenal glands, individuals subjected to emotional excitement will still show an increase in heart rate and a rise in blood pressure, but this is due to the release of noradrenaline from sympathetic nerve endings, which escapes into the general circulation to affect the heart and other structures. This re-action is much slower in onset and duration than that seen when the adrenal medulla is retained and the metabolic changes associated with adrenal medullary activation are minimal.

The close relationship between the adrenal medulla and the sympathetic system is readily explained. During embryonic development in the mammal, cells migrate from the neural crest to differentiate into sympathetic ganglion cells or into chromaffine tissue, so called from its reaction with chromates to produce a brown coloration. Paired masses of chromaffine tissue develop in the abdomen and become, when covered with layers of cortical cells, the adrenal medullae. Smaller masses of chromaffine tissue may also persist and are called paraganglia. Because of the common embryological origin of the cells of the adrenal medullae and sympathetic ganglia it is not surprising that the activities of the sympathetic division of the autonomic nervous system and that of the adrenal medulla are closely integrated, so that a discussion of the physiology of the adrenal medulla which neglects the parallel activities of the autonomic nervous system is somewhat unreal. Nevertheless, while the secretions of the adrenal medulla and sympathetic ganglia are similar, that of the medulla is released into the general circulation to affect a variety of targets, whereas that of the ganglia normally exerts a local action. Thus the adrenal medulla is a true endocrine gland lying under direct neural control.

Factors Influencing the Secretion of Adrenaline and Noradrenaline

The main secretory products of the adrenal medulla are adrenaline and nor-adrenaline. Noradrenaline is formed as a precursor of adrenaline during biosynthesis but the release of the two amines is separable under certain circumstances. There is evidence for the existence of two kinds of chromaffine cells in the adrenal medulla, one for the production of noradrenaline and the other for the methylated derivative. Upon appropriate treatment with iodate the noradrenaline secretory cells become darkly stained while those con-cerned with the production of adrenaline do not become tinted. It is argued that the cells can secrete independently of one another in response to different influences and that electrical stimulation of one region of the brain can favour the secretion of a particular hormone. Excitation of the nerves to the adrenal is said to release more adrenaline at a pulse frequency of 20 per second than at a frequency of less than 10 per second, but the gland may simply be respond-ing to the increased *number* of shocks applied, rather than to the increased frequency of stimulation. The output of sympathetic amines is maximal with a stimulation rate of 30–60 shocks/sec, but careful studies on the cat indicated that the proportion of adrenaline to noradrenaline did not vary with rate in any consistent way (Marley and Paton, 1961). It has not proved possible to separate fibres in the nerves controlling the output of one amine from those affecting another. Experiments involving electrical stimulation of the nerves have been considered unphysiological and the results attributed to a poor irrigation of the gland with blood (since vasoconstrictor fibres to the gland may also be activated) or to exhaustion by overlong stimulation of the nerve trunk (Malmejac, 1964). However, it is unlikely that exhaustion of the adrenal medulla occurs, for Marley and Paton (1961) showed that nicotine (which acts on the cholinergic synapses on the secretory cells) could still elicit a vigorous discharge of amines from an 'exhausted' gland. The effect probably stems from fatigue of the synapse (Marley and Prout, 1965). Nevertheless, some forms of emotional stimuli do increase the output of adrenaline while others affect that of noradrenaline more particularly (p. 46).

Marked species differences are apparent in the proportions of adrenaline and noradrenaline present in the adrenal gland. Noradrenaline forms but 2% of the amine content of the rabbit medulla, 9% of that of the rat, 12% in the hare, and some 17% in man. For the mouse the figure is 25%, the cat 41%, lion 55%, and whale 70–80% (von Euler, 1967). No satisfactory explanation is available for these differences. It has been noted that aggressive animals such as the cat and lion have large amounts of noradrenaline in the medulla (41% and 55%) whereas timid animals such as the rabbit, with 2%, have little. However, this generalization does not satisfactorily account for the peaceable nature of the whale.

The activities of the adrenal medulla are associated in large part in man and other species with the secretion of adrenaline, because adrenaline is present in the blood in larger quantity. This is in part a misconception, because most of the noradrenaline produced by the medulla is taken up by the heart, so reducing its relative concentration. Although it has been said that the bulk of the noradrenaline in the blood represents an overspill from the sympathetic nerves, most of the amine released at the nerve endings is taken up locally and so does not contribute to the pool in general circulation. It has proved difficult to determine the specific function of adrenaline, as opposed to noradrenaline, because of the overlap in actions. Only in a few instances has a different response to each amine been observed. Thus, noradrenaline causes constriction of the blood vessels in skeletal muscle whereas adrenaline causes dilation. Noradrenaline increases the peripheral resistance while adrenaline brings about a decrease, and noradrenaline does not alter the cardiac output, which is increased by adrenaline. Both amines mobilize fat and raise the plasma levels of free fatty acids and both raise the consumption of oxygen. But doses of noradrenaline which increase blood pressure by 50% hardly influence the oxygen consumption of man, whereas adrenaline has a marked effect. The hyperglycaemic and glycogenolytic responses to noradrenaline are 15–20% less than to adrenaline. To judge from the increased activity of the adrenal medulla that is apparent under stressful conditions the gland enhances the response of the sympathetic system. However, the endocrine organ cannot act as a substitute for the sympathetic system. Sympathectomized cats show fear and anger like the normal feline but the bodily reactions are depressed, with the rise in blood pressure and body temperature being minimal. When exposed to cold, sympathectomized cats cannot check loss of heat so that the body temperature falls more easily and they shiver more violently. Other noxious influences such as haemorrhage, anoxia, and hypoglycaemia are withstood less readily.

The fact that emotional stress increases adrenal medullary activity was firmly established in cats by Cannon and his associates (Cannon and Rosenblueth, 1937), who used, for example, a barking dog as the stimulus. Since this pioneer work the response to a variety of emotional states has been studied in human subjects and laboratory animals.

In man it is very clear that the excretion of adrenaline in the urine is increased by mental work, taking examinations, viewing exciting films, parachute jumping, space flight and the anticipation of space flight (von Euler, 1967; Mason, 1968a).

Efforts have been made to differentiate between those stimuli that raise the output of adrenaline and those enhancing noradrenaline secretion in the light of the suggestion of Funkenstein and his associates (Funkenstein, 1956), that the characteristic response of persons who respond to psychological stress by directing anger inwardly is the release of adrenaline, whereas that of individuals who direct anger outwardly in the discharge of noradrenaline. The occurrence of marked individual differences under conditions of athletic competition, or when flying, has made generalization difficult, but psychological studies have left the impression that active aggressive emotional states lead to an increased noradrenaline secretion, while tense, anxious but passive subjects predominantly produce adrenaline (Mason, 1968a). This view is reminiscent of the postulation of Goodall (1951), based on studies of the adrenal glands of a variety of African mammals, that aggressive animals have a higher concentration of noradrenaline than the typically non-aggressive species. An effective stimulus for adrenal activation need not involve personal threat although emotional arousal is required. Levi (1968) has described experiments in which a group of office girls viewed a series of films on several consecutive evenings. A fall in the output of adrenaline occurred during the screening of bland films of natural scenery and a marked rise was apparent when the tragic and agitating film 'Paths of Glory' was shown. The subjects reported feelings of anger and excitement during the film, whereas on the next evening the comic 'Charley's Aunt' elicited laughter and amusement. Nevertheless, the excretion of adrenaline again rose. In these experiments it seemed that the quality of emotional arousal, pleasant or unpleasant, was less important than the degree of arousal.

Some of the current uncertainty in the interpretation of experiments involving measurement of the urinary output of adrenaline and noradrenaline may stem from the use of excretion as an index of adrenal function, for it is almost impossible to establish a point-to-point relationship between the stimulus and the response of the medulla by this means. This becomes very clear when it is realized that the secretion of adrenaline can show a ten-fold increase in two minutes in man or monkey, and can fall back to normal almost as rapidly. From this point of view, the measurement of the plasma levels of the amines is much more reliable, although venipuncture alone causes a rise in blood level, and the occurrence of a diurnal rhythm in medullary function, with the production of amines rising abruptly on waking, needs also to be taken into account. Some of these objections can be overcome by working with monkeys equipped with an indwelling cardiac catheter through which blood can be withdrawn unbeknown to the animal. Mason, Mangan, Brady, Conrad, and Rioch (1961) found that moderate rises in the plasma concentration of noradrenaline occurred in response to a wide variety of stressful conditioning procedures and that comparable increases in the concentration of adrenaline were less frequent. The procedures which favoured noradrenaline release were familiar to the monkey and rather stereotyped, though unpleasant. Those promoting adrenaline secretion contained an element of uncertainty, unpredictability, and threat with the need for self-protection. On these occasions the output of noradrenaline and adrenocortical hormone also rose.

Alongside the physiological effects of the administration of adrenaline to human subjects, psychological changes have been described. Anxiety, excitement, tenseness, exhilaration, agitation, and fear have been reported but not all subjects react. Numerous studies have been cited by Schildkraut and Kety (1967), who point out that in many cases the response can be differentiated from a true emotion, with the subjects reporting that they felt 'as if' they were anxious. It is possible that adrenaline may produce a non-specific state of arousal with an accentuation of the prevailing mood, as is indicated by work in experimental situations designed to produce either euphoria or anger (Schachter and Singer, 1962). Since adrenaline occurs in the brain at a much lower concentration than noradrenaline, and is largely prevented from entry into the central nervous system by the blood-brain barrier, the question arises whether the psychological effects of the amine come from actions on the hypothalamus or through perception of the peripheral effects which mimic those normally experienced when anxious. The problem is by no means new and is often discussed in relation to the theories concerned with the psycho-physiological basis of emotion (Breggin, 1964; Grossman, 1967a). Noradrenaline is less psychically active than adrenaline upon injection. It is relevant that reserpine, a psychotropic drug that produces depression, seems to act by interfering with the binding of catecholamines within the neurones of the brain so that the store of monoamines becomes depleted. Administration of precursor amino acids that can cross the blood-brain barrier and raise the concentration of monoamines in the brain reverse reserpine-induced sedation in animals. Inhibition of the enzyme monoamine oxidase in the brain by treatment with iproniazid elevates the brain levels of noradrenaline and serotonin in most species, with the cat being a curious exception. Behavioural excitation in animals and the relief of depression in patients has been observed with the administration of iproniazid and has been attributed to the increased brain content of noradrenaline (Schildkraut and Kety, 1967).

Control of the Secretion of Adrenal Medullary Hormones

The secretion of hormones from the adrenal medulla of the adult is controlled through cholinergic fibres supplied from the thoracico-lumbar division of the sympathetic chain. Denervation of the adrenals by division of the splanchnic nerves abolishes the response to triggering stimuli but a residual and constant secretion of amines continues.

Electrical stimulation of the hypothalamus, midbrain, and certain areas of the cerebral cortex evokes secretion of medullary hormones (von Euler, 1967). Occasionally, as with stimulation of the orbital surface of the frontal lobe, the output of amines can be inhibited. There are indications that stimulation of some points in the hypothalamus favours the secretion of noradrenaline while that of others increases the release of adrenaline. However, these results have proved unacceptable to Goldfien and Ganong (1962) because of the use of indirect indices of catecholamine secretion or because shivering and changes in ventilation also occurred upon stimulation of the brain and may have caused secondary changes in amine release. In the work of Goldfien and Ganong (1962) stimulation of points in the dorsomedial and posterior nuclei of the hypothalamus which increased the secretion of adrenaline in dogs did not consistently produce a rise, or fall, in blood pressure, and activation of the adrenal medulla was not associated with release of adrenocortical hormones. On this basis the hypothalamic mechanisms controlling the adrenal medulla and adrenal cortex were separable.

A variety of endogenous stimuli can alter the release of adrenal medullary hormones. Changes in blood pressure are important, with a rise of pressure in the baroreceptor areas in the carotid artery or aortic arch bringing about a fall in hormone output, and a fall in pressure leading to greater hormone production. This reflex can be abolished by denervation of the carotid sinuses in vagotomized dogs and a seven-fold rise in amine secretion has been reported. Alongside this well defined reflex it has become clear that the electrical stimulation of afferent nerves in the anaesthetized cat can also increase the output of catecholamines (von Euler, 1967). The blood sugar level is also of great significance in governing medullary activity through the hypothalamus. Insulin hypoglycaemia in man increases the excretion of adrenaline, and alteration of the glucose level in the blood perfusing the head of the cat modifies the release of adrenaline from the adrenal. An increase in the sugar concentration reduced the output and a decrease had the opposite effect (Dunér, 1953). The injection of glucose solution directly into the hypothalamus also reduces medullary activity, whereas 2-deoxyglucose (a competitive inhibitor for glucose receptors) increases adrenaline secretion. Electrophysiological experiments involving the recording of the activity of single units in the hypothalamus indicate that numerous cells, especially in the lateral and posterior area of the hypothalamus, fire rapidly during a hypoglycaemia associated with an activated EEG and the appearance of signs of sympathetic discharge such as dilated pupils, tachycardia, and cutaneous vasoconstriction. These changes were readily reversible by the infusion of isotonic glucose solution but not isotonic saline or sucrose. As has been apparent in other work with single units, closely neighbouring cells were inhibited by a fall in blood glucose which caused the excitation of others (Crone and Silver, 1964). Less specific factors which increase the secretion of medullary hormones are exposure to cold, hypothermia, burns, haemorrhage, and asphyxia.

In view of the operation of feedback mechanisms in the neuroendocrine control of a variety of hormonal processes it is of interest that adrenaline can act to reduce its own secretion. Two mechanisms are operative, with one acting peripherally at the carotid sinus and the other centrally. Thus, the addition of adrenaline to fluid perfusing a carotid sinus at constant pressure caused a reduction in adrenomedullary activity, perhaps through a direct action on the muscle of fibres in the wall of the sinus. In parallel fashion, the addition of adrenaline to the fluid perfusing the head of a dog after the elimination of all reflex inputs also reduced adrenal activity. Noradrenaline acted like its methylated derivative in this regard (Malmejac, 1964).

5

Neurosecretion

Anatomists and physiologists were aware that some nerve cells in the brain appeared to be secretory long before interest in their function was aroused. Neurones possessing some features of gland cells were described in the ganglia of several invertebrate species towards the end of the last century, but the significance of this characteristic tended to be overlooked until 1928, when Scharrer clearly suggested that cells in the preoptic nuclei of a teleost fish (*Phoxinus laevis*) were neuroglandular in character and might be endocrine in function (see Gabe, 1966; Sloper, 1966). The supraoptic and paraventricular nuclear areas of a wide variety of species then came under examination and efforts were made to trace the mode of elaboration of the neurosecretory material and to delineate secretory cycles that could be correlated with one or other physiological process. An argument over the manner of secretion ensued, in which some considered that the stainable material moved up the axon to the nerve cell where it was consumed, while others adopted the Scharrer view that the substance was formed in the neurones of the brain nuclei and travelled down the axons for storage in nerve endings within the infundibular process.

Although the supraoptic and paraventricular nuclei are very prominent in the hypothalamus of mammals (chapter 2), it was not until the late 1940s and early 1950s that the relationship between the histological appearance of these nuclei and the function of the neurohypophysis came under close study. This was mainly due to the application of the chrome-alum haematoxylin stain of Gomori, originally developed to stain the different cell types in the pancreatic islets, to the neurohypophysis by Bargmann and his school, and the realization that the amount of stainable material in the supraoptic and paraventricular nuclei could be roughly correlated with the content of antidiuretic hormone in the neural lobe of the hypophysis. It was also established that the nuclear regions themselves contained material with posterior pituitary activity, and that neurosecretory material tended to accumulate on the proximal side of a cut in the pituitary stalk (Green, 1966b; Sloper, 1966). Although the chrome-alum haematoxylin stain has been that most widely applied to the study of neurosecretion, other techniques with a histochemical basis can be used. Of these, paraldehyde fuchsin and alcian blue have been employed successfully.

Three pairs of nuclei contribute to the hypothalamo-hypophysial neurosecretory system of vertebrates. These are the supraoptic, paraventricular, and (in birds) the tuberal nuclei. Fibres from the cells in these nuclei make up the hypothalamo-hypophysial tract which runs through the median eminence to the neural lobe of the hypophysis. In mammals, the system is concerned with the release of two neurohypophysial hormones, antidiuretic hormone or vasopressin, and oxytocin; it has also been suggested that it is involved in the secretion of the neurohumoral factors (releasing factors) controlling pars distalis activity.

Neurosecretory cells are specialized in that a 'glandular' function is emphasized, but they retain all the other morphological characteristics of neurones in that they possess dendrites, an axon, and Nissl substance. They also transmit electrical impulses like other nerve cells so that action potentials have been recorded from the supraoptic and paraventricular nuclei. The cells in these nuclei differ from other neurones in possessing membrane bound granules more than 1000 Å in diameter, and in liberating a product into the bloodstream that is sufficiently long-lived to exert an action on the kidney or mammary gland (Bargmann, 1966). However, other neurosecretory cells have granules smaller than 1000 Å so that the possession of granules alone is not adequate for the definition of such a cell. De Robertis (1964), for example, has placed all cells within the nervous system in this category. It is certainly hard to avoid describing those cells within the brain that produce catecholamines, serotonin, and histamine, as well as those elaborating the releasing factors, as neurosecretory, and on this basis it seems that additional groups of neurosecretory cells await discovery. Those producing the neurohumoral mediators are a case in point. Perhaps a more satisfactory alternative is to use the definition of Green (1966b), who regarded a neurosecretory cell as a neurone that liberated an active substance into the bloodstream or tissue fluids, thereby influencing the behaviour of somatic cells not necessarily in immediate contact with it.

Neurosecretory neurones could be important in the regulation of endocrine mechanisms in several ways (Bern, 1966). The neurones may secrete hormones that enter the blood to directly influence the activity of target organs, such as the kidney or uterus. They may produce substances which change the output of other endocrine glands, as the releasing factors act on the pars distalis through the hypophysial portal system. Alternatively, the neurosecretion may diffuse across membranes to affect adjacent endocrine tissue, as occurs for the pars intermedia in teleosts, or neurosecretory fibres may end directly upon endocrine cells, as in the pars intermedia of elasmobranchs.

The neurosecretory material stainable by the Gomori procedure is apparently synthesized by the endoplasmic reticulum, packaged by the Golgi apparatus of the cell body, and moves as membrane bound granules down the axon to the nerve ending. Neurosecretory material is particularly rich in cystine and when radioactive cystine is injected into the subarachnoid space in the rat it is selectively accumulated by the supraoptic and paraventricular nuclei. The time course of acquisition of the amino acid by the various parts of the neurohypophysis has been followed and cystine shown to be detectable in both nuclei 5 minutes after injection, with large amounts being present after 30 minutes. Since 10 hours elapsed before the labelled substance appeared in the pars nervosa, the concept of axonal transmission appears entirely reasonable (Bargmann, 1966; Sloper, 1966). The concept of axonal flow in neurosecretory and other nerve cells is supported by the accumulation of stainable neurosecretion and biological activity above transections of the pituitary stalk, but the mechanism producing the flow is not understood. Streaming of axoplasm has been observed in tissue cultures derived from the hypothalamus, as well as in other material of neural origin. Following division of the hypothalamo-hypophysial tract, or removal of the neural lobe, the stumps of the axons heal and accumulate neurosecretory material. This material can be released into the blood, and once this capacity is regained, the neurohypophysial deficit caused by the operation is relieved. It is also possible that the axons themselves can elaborate neurosecretion. Reorganization of the ends of the cut fibres occurs in time, so that after a month or so the stump of the pituitary stalk takes on the appearance of a regenerating neurohypophysis.

Neurosecretory activity within the hypothalamus is linked with that of the posterior lobe and overall is indicative of neurohypophysial function. When the demand for posterior lobe hormone is increased, as in thirst which causes secretion of antidiuretic hormone, the amount of neurosecretory material stored in the posterior lobe falls, and the depletion is approximately proportional to the intensity of the stimulus. Painful experiences and stresses also cause depletion of neurosecretory material from the neurohypophysis and this change may be related to the antidiuresis that is produced by a stress. The fall in content is detectable within a few minutes in the rat and the content is restored in a few hours. However, the visualized neurosecretion is not to be simply equated with the hormones of the neurohypophysis. Neurosecretory material can be extracted from unfixed samples of hypothalamus by organic solvents (it is not soluble after fixation), but although it retains the capacity to be stained it does not possess the activity of the neurohypophysial hormones. A polypeptide called neurophysin may act as a stainable, but separable, carrier of neurohypophysial hormones. Neurophysin does not possess biological activity in any high degree and is separable from oxytocin and vasopressin by precipitation with trichloracetic acid, by countercurrent distribution, or by electrodialysis. Reconstitution of the original complex is possible, and mammalian neurophysin can be used as a specific absorbent for the neurohypophysial hormones of birds, amphibia, and fish. Complexes in which synthetic analogues of posterior pituitary peptides are bound to the protein can also be formed (Acher, 1968). The application of molecular sieving and ion-exchange chromatography to neurophysin has recently led to the separation of two principal hormone-binding proteins (Hope and Hollenberg, 1968).

The precise mode of release of neurosecretory material at the nerve endings is not understood. The substance may be discharged by the acetylcholine which is also present in nerve endings, and may elicit a local breakdown of the terminal membrane. However, release of hormone has not occurred when isolated nerve endings (in the form of neural lobes) were treated with acetylcholine, although depolarization of the fibres in the isolated pars nervosa of rats by the application of a high concentration of potassium caused release of hormone. The entry of calcium ions into the secretory cells may be important in the liberation of the active peptides and may act by causing dissociation of the hormones from the complexes formed with neurophysin, or by promoting the extrusion of the contents of the neurosecretory particles from the endings (Ginsburg, 1968). Other possibilities include the breakdown of neurosecretory granules within the ending and diffusion of the hormone out of the axon, or the fusion of granules with the axonal membrane in the course of the release of their contents. Synthesis of hormones may begin in the cell bodies with the formation of biologically inactive precursors which are progressively converted to biologically active peptides as they move toward the nerve endings. Two kinds of membrane bound granules exist in the nerve endings in the neural lobe, with one exhibiting predominantly antidiuretic activity and the other being mainly oxytocic (La Bella, 1968). The granules are separable by differential centrifugation and this observation lends support to the suggestion that the neurosecretory system may be made up of two kinds of neurones, each secreting a different neurohypophysial hormone.

Neurosecretion has often been regarded as a source of the mediators involved in the control of the pars distalis. The association is highlighted by the changes in the amount of neurosecretory material in the hypothalamus that run in phase with the seasonal variation in sexual activity in birds, and the occurrence of ultrastructural signs of hyperactivity in the supraoptic and paraventricular nuclei of mammals after castration, and which disappeared upon replacement therapy with sex hormones (Zambrano and de Robertis, 1968b). It is possible that neurosecretion is released into the portal vessels of mammals and the suggestion takes on plausibility because the administration of posterior lobe hormones has been shown to cause ACTH, gonadotrophin, TSH, and growth hormone release. Neurosecretion accumulates in the median eminence in some species and light microscopic studies have indicated that branches of neurosecretory fibres end in close proximity to the primary plexus of portal vessels so that the short portal vessels could deliver neurohypophysial polypeptides in high concentration to the pars distalis. However, in work with the electron microscope typical neurosecretory axons have not been seen to make contact with the capillaries of the primary plexus of portal vessels (Zambrano and de Robertis, 1968a).

The relationship between the neurohypophysis and anterior pituitary activity has been discussed in detail by Martini (1966), where the assembled arguments for and against the identification of antidiuretic hormone as the chemotransmitter involved in the control of adrenocorticotrophic hormone (ACTH) secretion provide an excellent insight into the problems of workers in this field. In favour of the action of vasopressin is the close association between the release of both the anterior pituitary and posterior pituitary factors in response to noxious stimuli. Administration of adrenocorticoids has been reported to inhibit the release of both hormones, and to raise the content of stainable neurosecretion in the supraoptic-hypophysial system; while adrenalectomy reverses these changes. In some experiments the release of ACTH and the neurosecretory activity in the hypothalamus have not altered in parallel, but, in explanation, it is pointed out that the amount of stored neurosecretion reflects the balance between production and release and should not be taken to indicate the rate of discharge. On this basis, of course, all deductions made on the basis of study of histological sections of the hypothalamus are hazardous. Removal of the neural lobe, or dehydration, depresses the release of ACTH in response to stress, but the sensitivity is restored by treatment with antidiuretic hormone. This factor also causes ACTH secretion when given to the intact individual or when applied to cultures of anterior pituitary tissue. Those opposed to the concept have argued that antidiuretic hormone releases ACTH only when administered in pharmacological doses, for they are much higher than those necessary for antidiuresis, but, in turn, it is contended that the rapid clearance of neurohypophysial hormone from the blood must be taken into account and that it is the amount of hormone reaching the pars distalis that matters. Certainly, when antidiuretic hormone is infused into the arterial supply of the anterior pituitary gland very small amounts are effective, and it has been calculated that the concentration of polypeptide in the blood reaching the neural lobe may be relatively high. What is not clear is how much reaches the pars distalis. Paradoxically, diabetes insipidus, due to a lack of antidiuretic hormone, is not invariably associated with an impairment of the secretion of ACTH and this has posed a problem. In some studies of animals subjected to hypothalamic lesions or to removal of the posterior lobe, the adrenal glands and the plasma level of adrenal corticoids have been greater than in controls. Martini (1966) argues that sufficient antidiuretic hormone may nevertheless be released into the portal vessels to cause ACTH release, although not enough escapes into the systemic circulation to affect the kidneys. In order to eliminate this possibility experimentally it would be necessary to destroy all the neurones carrying antidiuretic hormone and neurosecretory material to the median eminence.

Although there is no doubt that the antidiuretic hormone can elicit ACTH secretion, few investigators believe that antidiuretic hormone is, or acts as, the neurotransmitter for ACTH, largely because chemical studies of these agents indicate that corticotrophin releasing and antidiuretic activity are separable (p. 74). Thus the physiological significance of the response of the pars distalis to antidiuretic hormone awaits resolution, but it may be that the neurohypophysial polypeptide is effective because it contains sequences of amino acids in the molecule which are also possessed by the true corticotrophin releasing factor.

6

Control of the Secretion of Neurohypophysial Hormones

The activities of the neurohypophysis have been associated with the control of kidney function, with uterine motility and the lactating mammary gland ever since pharmacological studies with extracts of the neural lobe of the hypophysis were undertaken at the turn of the century. Purification of a crude extract and separation of the material into two fractions with differing effects was achieved by Kamm, Aldrich, Grote, Rowe, and Bugbee in 1928, but a further quarter of a century elapsed before the structures of the hormones were determined (see Acher, 1966). It is now becoming accepted that the neurohypophysis secretes two separate factors, as opposed to a single factor with multiple activities which can be fragmented artificially; with one acting largely upon the kidney and the other being concerned with the function of the accessory sexual organs of the female.

As previously related, the production of neurohypophysial hormones is closely associated with the processes of neurosecretion, but it is admittedly difficult to link the production of neurosecretory material to the elaboration or release of one or other factor, although under extreme conditions a correlation between a fall in the content of neurosecretion and the discharge of hormone has been found. On the other hand, it is now very clear that the neural lobe of the hypophysis serves mainly as a storage organ for hormones formed within the hypothalamus. Damage to the supraoptic-hypophysial tract can block the movement of hormone down the pituitary stalk to the posterior lobe so that it is no longer surprising that the hypothalamus of the hypophysectomized dog contains more neurohypophysial hormone than that of the intact control. It is of interest that with the passage of time after transection of the pituitary stalk, or hypophysectomy, the stump of the stalk above the cut undergoes reorganization and comes to resemble a reformed pars nervosa. A substantial degree of neurohypophysial function is simultaneously restored.

The two hormones of the neurohypophysis are vasopressin and oxytocin. Although the two terms are generally accepted, they are not entirely satisfactory since the prime activity of the former is antidiuretic, not vasomotor, and oxytocin has a marked influence on the expulsion of milk from the mammary gland besides affecting the motility of the uterus.

Vasopressin is of doubtful physiological significance in the control of vasomotor activity (Pickford, 1966a) but is important in the control of water resorption by the kidney (Pickford, 1966b). In the absence of vasopressin the condition of diabetes insipidus ensues in which large volumes of urine of very low specific gravity are excreted. In order to compensate for the loss of fluid the diabetic individual must drink large volumes of water and becomes severely dehydrated if this is not available. The hormone increases the absorption of water by the distal convoluted tubules and collecting ducts and is active in minute quantity. An intravenous injection of 2–3 mU, approximately 2 nanograms, into a hydrated dog immediately increases the osmolarity of the urine and reduces the rate of formation. In man, the rate of urine flow can be reduced by an intravenous injection of 1 mU vasopressin. Because kidney function is affected by so little hormone it is not surprising that physiological stimuli causing the release of vasopressin from the neurohypophysis do not usually deplete the stock of the factor.

The activities of oxytocin cover a broader spectrum than those of vasopressin and are closely connected with reproductive processes in the female. Oxytocin causes contraction of the uterus and it has been suggested (Harris, 1947) that release of the hormone during coitus may be of value in accelerating the movement of spermatozoa up the genital tract toward the ovary. While there is no doubt that some oxytocin is released upon sexual stimulation, as is shown by the occurrence of milk ejection with orgasm in women, or after mechanical stimulation of the vagina and vulva in domestic animals, the importance of this reflex for sperm transport has not been established (Fitzpatrick, 1966). On the other hand, it is generally accepted that oxytocin is released at the end of pregnancy, during parturition. Delivery is promptly initiated in rabbits on the last day of pregnancy by the injection of oxytocin and the hormone has been used clinically to assist delivery for some sixty years. The pattern of uterine contractions caused by the hormone in women is very similar to that occurring during spontaneous labour. Milk ejection has been observed at delivery in women and domestic animals, and an oxytocin-like substance has been detected in the blood of sheep, goats, cows, horses, and women during labour. However, damage to the neurohypophysis does not necessarily prevent delivery, and normal labour has been observed in women with diabetes insipidus. Studies of hypophysectomized animals have given discordant results but dystocia has been commonly observed. In the absence of oxytocin, 'spontaneous' contractions of the uterus aided by reflex contractions of the abdominal muscles may suffice. Spinal anaesthesia, which blocks the abdominal straining reflexes, has been found to cause retention of the young at the portals of the genital tract with the last member of the litter frequently being retained. Milk ejection as a response to oxytocin has been mentioned on several occasions. This is the process by which milk is forced from the alveoli of the mammary gland into the ducts and sinuses for removal during suckling by the contraction of myoepithelial cells, which are sensitive to oxytocin and are strategically placed around the alveoli. In the absence of milk ejection, or the 'draught', or 'let-down' as it is also called, the young of many species are able to obtain but a fraction of the milk present in the mammary gland and die of starvation. Damage to the neurohypophysis which blocks the release of oxytocin disrupts lactation in the rat or rabbit, by preventing egress of milk, just as surely as damage to the anterior lobe of the pituitary which stops the secretion of prolactin. Oxytocin may be of less importance in ruminants, such as the goat and sheep, in which large cisterns provide storage space for the milk as it is produced (Linzell, 1963).

The composition and structure of vasopressin and oxytocin are known. Both hormones are polypeptides of relatively low molecular weight. The composition of oxytocin (mol.wt. 1007) appears to remain constant from species to species but two forms of vasopressin are known in mammals. One, arginine vasopressin, is found in the glands of man, cattle, horse, sheep, and most mammals, and the other, lysine vasopressin, is found in the pig and some other pig-like mammals (Heller, 1966). The close similarity in structure of the neurohypophysial hormones is illustrated by a comparison of the sequence of amino acids in each:

Arginine vasopressin	CyS. Tyr. Phe. Glu (NH₂). Asp (NH₂). Cys. Pro. Arg. Gly (NH₂)
Lysine vasopressin	CyS. Tyr. Phe. Glu (NH₂). Asp (NH₂). CyS. Pro. *Lys.* Gly (NH₂)
Oxytocin	CyS. Tyr. *Ile.* Glu (NH₂). Asp (NH₂). CyS. Pro. *Leu.* Gly (NH₂)

In view of the relatively minor differences in amino acid composition it is entirely reasonable that vasopressin exhibits oxytocic activity, albeit to a minor degree, and that oxytocin can affect water reabsorption to a small extent. Vasopressin possesses only one-sixth of the potency of oxytocin in causing milk ejection, while the antidiuretic activity of oxytocin is less than one-two-hundredth of that of its companion hormone.

In view of the overlap in activity of the two neurohypophysial factors it may be useful at this point to assemble some of the evidence for the separate release of vasopressin and oxytocin. The human response appears to be well defined in that osmotic stimuli which elicit the secretion of vasopressin do not discharge detectable quantities of oxytocin. Conversely, stimulation of the cervix or mammary glands in women can cause oxytocin secretion without any antidiuretic effect, and the emotional stimuli which can block the reflex release of oxytocin (p. 61) increase the output of vasopressin, as is established by the consequent antidiuretic action. A patient described recently, suffered from diabetes insipidus and was unable to produce concentrated urine in the absence of administered vasopressin. Nevertheless, the reflex release of oxytocin and milk ejection occurred upon suckling, and even when nursing was anticipated (Chau, Fitzpatrick, and Jamieson, 1969). In experimental animals, it has been found that lesions in the infundibular region of the hypothalamus may cause diabetes insipidus, although normal parturition and milk ejection can still occur. Other lesions can result in a disturbance of parturition and of milk ejection in the absence of diabetes insipidus. In cats, electrical stimulation of the brain in the vicinity of the paraventricular nuclei released oxytocin without the associated discharge of any detectable vasopressin, whereas stimulation of the region around the supraoptic nuclei caused vasopressin secretion without detectable release of oxytocin (Rothballer, 1966); the changes observed in the secretion of these polypeptides by others are described later. Haemorrhage, in moderate degree, causes the release of vasopressin in the absence of detectable amounts of oxytocin in the cat and rat (Fabian, Forsling, Jones, and Lee, 1969). It may also be relevant that diabetes insipidus is known as an heritable disease in rats and occurs because the hypothalamic neurones are unable to form vasopressin, although oxytocin is synthesized normally (Valtin, Schroeder, Benirschke, and Sokol, 1962). Because these rats are much more sensitive to vasopressin and oxytocin than normal animals, they have been employed with benefit in the assay of threshold amounts of the polypeptides.

Control of Vasopressin Secretion

The secretion of vasopressin is most readily affected by changes in the composition of the blood, by haemoconcentration or the consequent rise in osmotic pressure. Emotion is also influential, as demonstrated by the reduction in urine flow that occurs when a dog is shown a cat. If the emotional stimulus is severe and stressful, the adrenaline that is also discharged causes an immediate but short-lived antidiuresis that is followed by the more prolonged effect due to vasopressin. Pain causes a similar reaction that is seemingly specific because it can occur in the absence of changes in blood osmolarity, volume, or pressure, can be elicited by stimulation of a sensory nerve in an anaesthetized dog, and is prevented by removal of the infundibular process. Antidiuresis also accompanies the induction of ischaemic pain in man (Heller and Ginsburg, 1966). The secretion of vasopressin can be readily conditioned in dogs by combining a tone with a nociceptive stimulus such as electric shock to the foot. After a series of trials, antidiuresis occurs with the tone alone (Corson, 1966).

In man, deprivation of water overnight elevates the plasma level of antidiuretic hormone, as does more severe dehydration in the rat. The thirsting individual suffers a decrease in plasma volume and an increase in plasma osmolarity. An artificial increase in the osmotic pressure of the plasma reaching the brain by the intracarotid infusion of hypertonic saline in conscious dogs readily caused the liberation of vasopressin, as shown by the occurrence of antidiuresis (Verney, 1947) and the appearance of a measurable quantity of the hormone in the blood of the jugular vein (Ames, Moore, and van Dyke, 1950). Repetition of such experiments with the progressive ligation of branches of the arterial supply to the brain indicated that the receptors (osmoreceptors) were located within the hypothalamus or preoptic area (Jewell and Verney, 1957). The supraoptic nuclei are particularly richly supplied with capillaries (Liss, 1965/66) and the electrical activity of neurones in or around this region alters after the intracarotid injection of hypertonic saline. However, supraoptic neurones respond to other stimuli besides osmotic, and the rate of discharge of cells in parts of the brain outside the hypothalamus, such as the medullary reticular formation near the area postrema, is also increased upon injection of hypertonic solution. However, the secretion of antidiuretic hormone is not below normal in cats in which the hypothalamus has been isolated from the rest of the brain (Woods and Bard, 1960).

Evidence for the occurrence of osmoreceptors outside the brain is also available (Haberich, 1968). These are believed to be located in the portal circulation, for infusion of water into the portal veins of rats at a rate which altered the osmolarity of the portal blood by less than 1% caused a marked diuresis. Comparable infusions into the inferior vena cava did not bring about this result. It can be argued that the location of osmoreceptors in the liver is entirely reasonable, for the osmotic concentration of the portal blood varies considerably with changes in intestinal absorption and secretion. After the ingestion of water a shift of electrolytes, mainly sodium, chloride, and bicarbonate from the splanchnic circulation into the water in the lumen of the intestine occurs to produce an isotonic solution. Consequently, the osmolarity of the portal blood falls, and in response to this osmotic stimulus the liver increases its water content considerably. The liver may thus act as a buffer which prevents or delays a decrease in the osmolarity of the systemic blood. Since the osmotic responses from the liver could be blocked by cutting the small branch of the vagal nerve which runs from the stomach to the liver, the vagus may provide the afferent pathway to the brain.

It has often been suggested that a fall in blood volume brings about secretion of antidiuretic hormone, just as an infusion of blood, which does not significantly raise arterial pressure, induces a diuresis which lasts until urine equivalent to the volume infused is excreted (Gauer and Henry, 1963; Sawyer and Mills, 1966). Such a change may be monitored through an alteration in the volume of blood contained in the thorax which acts through receptors in the atria of the heart supplied by the vagus nerves. As the atrial receptors become activated by an increased cardiac return, the secretion of vasopressin would be inhibited. This view is supported by experimental studies in dogs, where the inflation of balloons in the left atrium led to a diuretic response on the part of the kidney. According to Gauer (1968), the outstanding importance of the stretch receptors in the low pressure venous system lies in the fact that they can sense minute changes in blood volume that are too small to affect arterial haemodynamics or intra-arterial receptors. Thus, elevation of the left atrial transmural pressure by as little as 2–7 cm water has resulted in a significant reduction in blood antidiuretic hormone concentration. The increase in the plasma level of this hormone after a sudden reduction in the pressure within the carotid sinus, in the atria, or after haemorrhage, pain, or surgical stress, has been found to be greater than that produced by water deprivation. A different experimental approach has been adopted in human physiological studies and an increased intrathoracic blood volume achieved by applying artificial respiration under negative pressure. Again a water diuresis ensued, except when hypertonic saline was infused to raise the plasma osmolarity. Under conditions of positive pressure ventilation antidiuresis was induced in hydrated men, when the opposite effect would normally be anticipated, and could be prevented by the ingestion of ethanol. The results are satisfactorily accounted for by changes in the secretion of vasopressin, although it has been found that a reduction in the intrathoracic blood volume in water loaded men by immobilization or tilting produced an antidiuresis only after the subjects fainted and had experienced a fall in blood pressure. In dogs, the activity of atrial receptors, as recorded from the vagus nerves, was reduced by haemorrhage even before there was a fall in blood pressure. Division of the vagus nerves in the neck has been reported to increase the plasma concentration of vasopressin. In other experiments, stimulation of the central end of the vagus in dogs raised the blood pressure or inhibited a water diuresis, and these results have been taken to indicate that there are afferent fibres in the vagus which elicit vasopressin release, alongside others which are inhibitory. The importance of atrial receptors in the secretion of vasopressin after haemorrhage is the subject of dispute, and attention has been directed to the decline in blood pressure that also occurs (O'Connor, 1962; Heller and Ginsburg, 1966). The results of a recent re-examination of the effects of haemorrhage in the cat indicated that vasopressin was released independently of oxytocin and that secretion was more closely related to the fall in blood pressure than to changes in blood volume. A rise in the concentration of vasopressin in the blood was also observed after stimulation of the peripheral end of the cut vagus in cats that were not subjected to major bleeding, but a fall in blood pressure of about 80 mm Hg was necessary to bring about this effect (Beleslin, Bisset, Haldar, and Polak, 1967). The mechanisms controlling vasopressin release in the dog appear to be more sensitive in that a fall in arterial blood pressure of 25 mm Hg induced by haemorrhage was sufficient to evoke some discharge (Rocha e Silva and Rosenberg, 1969).

Vasopressin release occurs after stimulation of a variety of structures within the brain (Rothballer, 1966; Sawyer and Mills, 1966). Excitation of ascending pathways in the medulla and pons and many points in the lateral reticular formation are effective, as is stimulation of points in the tegmentum and periaqueductal grey matter of the midbrain which receive collaterals from the ascending spinal tracts, are traversed by ascending reticular pathways, and project to the hypothalamus. In turn, stimulation of the hypothalamus readily induces antidiuresis or a rise in blood pressure. Many efforts have been made to associate the release of vasopressin with one or other of the magnocellular nuclei; the clearest results have been obtained in the cat after division of the spinal cord in the neck to suppress activation of the sympathetic system and adrenal medulla (Rothballer, 1966). When the electrodes were placed near the supraoptic nuclei, vasopressin was released without detectable amounts of oxytocin. A similar finding was obtained by Bisset, Hilton, and Poisner (1967). In their experiments, excitation of the supraoptic nucleus, paraventricular nucleus, or the supraoptic-hypophysial tract caused the release of high concentrations of vasopressin with extremely little or no oxytocin. Significant quantities of oxytocin were released only after stimulation of the tuberal region of the hypothalamus. Study of the hormonal content of extracts of parts of the hypothalamus established that both vasopressin and oxytocin were

59

present in the supraoptic and paraventricular nuclei, but the origin of the neurones releasing oxytocin was not clearly revealed. These results differ from others derived from work on the rabbit, where stimulation of the paraventricular nuclei was most effective in causing an increase in intramammary pressure, while stimulation of the supraoptic nuclei was followed by prominent pressor and milk ejection responses. Maximal neurohypophysial hormone release was achieved by stimulation of the hippocampal rudiment, central grey, subthalamus and lateral hypothalamus, but both factors were liberated except in the case of the hippocampal rudiment and nucleus accumbens, where only oxytocin secretion was affected (Woods, Holland, and Powell, 1969). The paraventricular nuclei have been found to contain much more oxytocin than vasopressin in all species studied. Electrical stimulation of these nuclei in the goat has caused antidiuresis, while excitation of the supraoptic nuclei has elicited milk ejection as well as an inhibition of a water diuresis (Heller and Ginsburg, 1966).

A cholinergic mechanism may be involved in the release of vasopressin from the neurohypophysis, for the injection of acetylcholine into the supraoptic area of the dog caused antidiuresis, as did the application of some anticholinesterases (Pickford, 1947). These results can be accounted for in two ways, for a cholinergic synapse could be employed by afferent neurones connected with the cells of the supraoptic nuclei, or acetylcholine might be involved in the release of hormone from the nerve endings in the neurohypophysis. It may be through this cholinergic mechanism that nicotine administration, whether by injection or through cigarette smoking, promotes the release of vasopressin, although a variety of hypnotic drugs act likewise. On the other hand, ethanol suppresses the release of antidiuretic hormone, except in the case of the discharge evoked by haemorrhage (Heller and Ginsburg, 1966).

Control of Oxytocin Secretion

Stimuli from the genitalia and mammary glands, as well as psychic stimuli, are important in the control of the release of oxytocin. Manipulation of the cervix in the cow and goat causes milk ejection and the importance of a humoral link in this process has been elegantly demonstrated by cross-circulation experiments in the sheep, where vaginal distension in a donor animal caused milk ejection in a lactating recipient ewe (Debackere, Peeters, and Tuyttens, 1961). The milk ejection response was comparable to that caused by an injection of oxytocin. In rabbits, with a bicornuate uterus, distension of the cervix or of one horn increased the motility of the other. The reflex was not solely mediated by the nervous system because it showed a long latency, could be reproduced by the administration of oxytocin, and was abolished by destruction of the pituitary stalk (Ferguson, 1941). Despite the evidence for the secretion of oxytocin during labour, it is not agreed that parturition is precipitated by a sudden release of oxytocin. Nor are the causative factors promoting the discharge of oxytocin fully understood (Cross, 1966; Fitzpatrick, 1966). In the eyes of some, the activities of progesterone loom large (Csapo and Wood, 1968). Progesterone is known to increase the membrane potential of the cells of the myometrium and so desensitize the uterus to oxytocin. A fall in the plasma level of progesterone at the end of pregnancy would progressively diminish the progesterone block to myometrial contraction and favour the action of oxytocin, so prompting delivery. It is also possible that progesterone depresses the response of hypothalamic neurones to stimuli from the genital tract and that as the influence of the steroid wanes, an increased secretion of oxytocin ensues which assists delivery (Cross, 1966). However, these considerations appear to apply mainly to the rabbit and rat, and not to women, sheep, and guinea-pigs, where progesterone administration, even in toxic dosage, fails to prolong pregnancy beyond its normal term. Measurement of the plasma levels of oxytocin in sheep have shown that distension of the vagina in cycling or lactating ewes elicits the secretion of the hormone, but also that in pregnant animals the concentration of oxytocin falls (Roberts and Share, 1968). Further, the plasma level of progesterone may not fall until delivery is well underway and separation of the placenta from the uterus has occurred. By that time it would be too late to influence the secretion of oxytocin, except to support the evacuation of the uterus. The human uterus responds to oxytocin well before the end of pregnancy.

Because of the ease with which the ducts in the teats of experimental animals, and women, can be cannulated, or the progress of lactation followed, the neural basis of the milk ejection reflex has been studied in some detail. Afferent impulses from the nipples or teats are probably initiated through the somatic sensory innervation. Denervation of the nipples disrupts lactation in the rat, as does interruption of the projection through the spinal cord. The pathway in the spinal cord appears to lie ipsilaterally deep within the lateral funiculus but has not been traced through the brain stem to the hypothalamus.

Milk ejection has followed electrical stimulation of a number of areas in the brain (see Rothballer, 1966; Tindal, 1967; Aulsebrook and Holland, 1969a). These include the medial lemniscus, reticular formation, subthalamus, supramamillary area, mamillary peduncle, septum, fimbria of the fornix, piriform cortex, hippocampus, and amygdala. It is unlikely that all the structures contribute to the ascending path employed by the suckling stimulus, and several, such as the hippocampus and amygdala, may supply a collateral input. The optimal site in the midbrain of the guinea-pig for the release of oxytocin lay in the far lateral tegmentum, close to the lateral lemniscus and medial geniculate body, and stimulation of this area is known to facilitate the electrical activity transmitted from the amygdala to the ventromedial nuclei of the hypothalamus. Lesions in the amygdala block milk ejection in the rat, but the significance of correlations between the existence of connections between brain structures and the effects of lesions, or stimulation, is not understood.

The milk ejection reflex is readily conditioned. In cattle, 'let-down' has been shown to occur as soon as the processes associated with milking on the farm have been started and the rattling of buckets or the sight of a calf may suffice. Conditioning is not confined to cows or goats and has been observed many times in women. Psychic stimuli, besides favouring milk ejection, can also block the response: fright or pain will prevent 'let-down' in cattle, with embarrassment or discomfort being additional factors in women. Two processes are concerned in the inhibition of milk ejection, for both peripheral and centrally operating mechanisms have been studied. Peripherally, the vasoconstriction following the sympathetic activation associated with stress (p. 45) limits access to the myoepithelial cells of any oxytocin released by the neurohypophysis. Adrenaline injected into rabbits or dogs immediately before nursing has blocked milk ejection. On other occasions, an inhibition of milk ejection can be produced which can be overcome by the administration of oxytocin. Painless restraint of nursing rabbits or of guinea-pigs is effective and this form of blockade can be weakened by treatment with tranquillizing drugs. Anaesthetics, and ethanol, also inhibit oxytocin release by a central action. A search for the central inhibitory system in lactating rabbits has been made by Aulsebrook and Holland (1969b). They first located points in the brain through which oxytocin release could be induced by electrical stimulation, and then examined the effect of prior stimulation of other regions on the expected milk ejection. The release of oxytocin following stimulation of the paraventricular or supraoptic nuclei could not be inhibited by stimulation of points which had been shown to be inhibitory in other observations, so that the central inhibition of oxytocin secretion would appear to be exerted upon the afferent pathways to these nuclei, although it is possible that the electrical stimulus over-rode any direct inhibition of the neurones. Inhibition could be elicited by stimulation of a variety of structures, including the superior colliculi, dorsomedial nucleus of the thalamus, and the piriform cortex, and appears to be related to the more ventral and lateral portions of the limbic system, which may be particularly concerned with self-preservation.

There is little question that the excitatory fibres concerned with the control of oxytocin secretion converge upon the hypothalamus. Within the diencephalon, oxytocin release has followed electrical stimulation of either the supraoptic or paraventricular nuclei. After a detailed examination of the effect of lesions confined to the paraventricular nuclei on the hormone content of the infundibular process, Olivecrona (1957) concluded that the paraventricular nuclei were more concerned with oxytocin secretion, and the supraoptic nuclei with vasopressin. Destruction of the paraventricular nuclei in pregnant cats, without damage to the supraoptic nuclei, caused dystocia and depressed the oxytocin content of the infundibular process (Nibbelink, 1961). In the rabbit, stimulation of the paraventricular nuclei has caused the release only of oxytocin, whereas both oxytocin and vasopressin were released after excitation of the supraoptic nuclei (Aulsebrook and Holland, 1969a). From the information presented earlier (p. 59) it will be evident that the results of different workers on the effects of stimulation of the neurohypophysial nuclei are often discordant. This could be due to over-intense stimulation, with consequent involvement of structures located at a distance from the electrode, or, in the case of the supraoptic nuclei, to the stimulation of fibres of the paraventricular nuclei which closely approach the basal nuclei before joining the supraoptic-hypophysial tract. It may also be noted that the neurones of the supraoptic nuclei are rather spread out over the optic tract so that excitation of a large number is less practicable than in the case of the paraventricular nuclei. The fact that both polypeptides are released from stimulation of the supraoptic nuclei then becomes more striking.

While oxytocin may be the predominant polypeptide in the paraventricular nuclei, and vasopressin present in larger amount in the supraoptic nuclei, it now seems that there is an overlap in function. Stimulation of the supraoptic-hypophysial tract causes milk ejection and the reflex is blocked by interruption of the tract. It is curious that the intracarotid injection of hypertonic saline has been found to cause oxytocin release in the goat, dog, and rabbit, presumably through the mediation of osmoreceptors, and that a similar response has been observed in rabbits with hypothalamic islands (Sundsten and Sawyer, 1961). Evoked potentials have been recorded in the supraoptic nuclei of the cat after electrical stimulation of the uterus (Abrahams, Langworth, and Theobald, 1964).

7

Control of the Secretion of Adrenocorticotrophic Hormone

Under normal circumstances the adrenal glands secrete a spectrum of steroids which are commonly divided into three groups, the glucocorticoids, the adrenal androgens, and a mineralocorticoid. Of these, the last is most important in maintaining life after adrenalectomy, though it should be remembered that adrenalectomized rats and mice can be maintained in good health provided sufficient sodium chloride is given in the diet. In cases of adrenal insufficiency in man a variety of symptoms become manifest. These include a loss of appetite, vomiting, muscular weakness, hypotension, and a fall in basal metabolic rate. Conversely, when the adrenal cortex secretes excessive quantities of hormone, or when too much cortisone is given, hyperglycaemia and osteoporosis may be observed. Alongside these somatic changes, the psychic state may alter toward euphoria or, in the reverse direction, toward psychotic conditions. Such alterations in mental outlook may be regarded as indicative of an action of the adrenal hormones upon the brain.

The effects of adrenocorticotrophin (ACTH) include stimulation of the growth of the adrenal cortex through proliferation of the cells in the zona fasciculata and zona reticularis, and acceleration of the synthesis and release of steroid hormones of the glucocorticoid group. The main glucocorticoids secreted vary from species to species, being cortisol or hydrocortisone in man, dog, and guinea-pig, and corticosterone in the cat, rat, and mouse. Under the influence of ACTH, alongside the increased production of adrenal steroids, the concentration of cholesterol and ascorbic acid in the gland falls. The loss of cholesterol is understandable, since this substance could provide a starting material for the synthesis of steroid hormones (though the fall in cholesterol cannot be fully accounted for on this basis) but the depletion of ascorbic acid is less easily explained. Nevertheless, the fall in ascorbic acid concentration produced by ACTH has provided the basis of a well-established and popular assay for the pituitary hormone (Sayers, Sayers, and Woodbury, 1948). Adrenocorticotrophic hormone is known to induce a number of changes outside the adrenal gland. It stimulates the melanocytes in the skin to produce a bronzed appearance, which is characteristic of patients suffering from an excessive secretion of the hormone (this effect has also been attributed to the action of melanocyte stimulating hormone from the intermediate lobe), and can promote the breakdown of fat stored in adipose tissue.

Adrenocorticotrophic hormone is the only product of the pars distalis for which the chemical structure is known. The porcine type of hormone was totally synthesized by Schwyzer and Sieber in 1963 and is a relatively small polypeptide of some 39 amino acid residues. The half life of ACTH is extremely short, for the injection of this pituitary factor into rats, with study of the subsequent changes in blood level of hormone, has produced results indicative of a 50% loss in about 5 minutes, whereas studies of the half life of the endogenous hormone secreted in adrenalectomized rats, indicate that in this situation the half life is about 30 seconds. The half life of exogenous ACTH in man is believed to be of the order of 10 minutes (Evans, Sparks, and Dixon, 1966). This information is important, for it indicates that there must be a constant secretion in order to maintain measurable levels of the hormone in the blood. When the short half life of ACTH is taken into account it is not surprising that the action of the hormone on the adrenal gland is extremely rapid. An increased release of steroids into the adrenal venous blood can be detected two minutes after an intravenous injection of the trophin and thereafter the output can be raised three to five times above the basal level. The adrenal glands of freshly hypophysectomized animals are much more responsive to ACTH than those of intact individuals, but the responsiveness rapidly declines and is below normal 24 hours after the operation.

Study of the neural control of adrenal function began at a time when only indirect responses to ACTH could be used to follow the experimentally induced changes in the output of this hormone. It was not possible to trace the changing secretion of adrenal corticoids so that the effects studied (fall in the number of eosinophils or lymphocytes per unit volume of blood, decrease in size of the thymus gland), were several times removed from the pituitary controlling mechanism. In retrospect, it is remarkable that so much useful information accrued from the use of such crude indices. However, the curious effect of the pituitary hormone in reducing the ascorbic acid content of the adrenal gland proved to be of major utility. The adrenal cortex normally contains the highest concentration of ascorbic acid in the body but this abruptly declines upon treatment with ACTH in assay animals. The adrenal ascorbic acid content is not depressed by ACTH in all species (the cat and dog are exceptions) and the significance of the change is poorly understood. Nevertheless, an assay based on the fall in adrenal ascorbic acid brought about by ACTH in hypophysectomized rats proved to be of the greatest value in research on the physiology of this hormone, although advances in steroid chemistry have led to the displacement of this procedure in favour of others based on steroidogenesis within the adrenal gland.

The Environment and ACTH Secretion

Research in this field has been markedly influenced by the concept of 'stress', as enunciated by Hans Selye (1936, 1950). This investigator early pointed out that the release of the adrenal stimulating hormone from the anterior pituitary gland was induced by a great variety of stimuli originating from adverse environmental conditions and developed the view that naturally occurring stresses can cause 'diseases of adaptation' or a 'general adaptation syndrome' due to prolonged activation of adrenal function. It is now very clear that the secretion of ACTH fluctuates in response to a wide variety of stimuli. These range from psychological conditions generating anxiety, fear, and anger, to exposure to cold, heat, or surgical trauma. Disturbance of almost any kind is sufficient to increase the output of this trophic hormone, so that a rise in the blood level of adrenal steroids has been traced in emotionally disturbed patients, relatives accompanying patients suddenly taken to the casualty department of a hospital, students examined just before important examinations, and pilots flying jet aircraft (Mason, 1968b).

Because ACTH secretion is so labile, the question has arisen whether the release of this hormone continues in the absence of provocation, but a satisfactory answer has not been obtained. Rupture of the pituitary stalk, either by surgical stalk section or by transplantation of the gland, depresses the release of ACTH, although the decline is not as marked as after hypophysectomy. This may be due to the fact that the hypothalamic releasing factor for this hormone (p. 72) can reach the isolated gland through the systemic circulation.

Study of the effects of environmental changes has been carried over from laboratory colonies to mammalian populations in general in an attempt to explain some of the spectacular fluctuations in the numbers of certain species. Because of the greater strife which occurs in association with increased numbers in a colony mortality increases, reproductive capacity declines, and the population pressure is relieved. The adrenal gland may be important in the mediation of this response. Christian suggested in 1950 (Christian and Davies, 1964) that stimulation of pituitary-adrenocortical activity and inhibition of reproductive function would occur with increased population density, for increased adrenocortical secretion would increase mortality indirectly through lowering the resistance to disease, to parasitism, and to other adverse conditions. In experiments with laboratory mice, progressive adrenocortical hypertrophy and thymic involution occurred with increasing population density. Somatic growth was suppressed and reproductive function curtailed in both sexes; sexual maturation was delayed or even inhibited at high population densities. It would be wrong to attribute all these effects to an enhancement of ACTH release but this is unquestionably important. The level of adrenocortical activity can be directly related to social rank in rats and dogs, while the adrenals from mice kept in groups produce more corticosteroids *in vitro* than those from singly caged animals. Comparable increases in the plasma levels of the adrenal steroids of rats maintained in colonies, above those of animals kept in groups of four, have also been measured. It is also interesting to find that high-ranking individuals generally do not have enlarged adrenals, in contrast to their low-ranking subordinates (Christian and Davies, 1964).

In the absence of stressful stimuli a diurnal variation in the output of ACTH becomes apparent, with peak discharge of adrenal hormones occurring during the morning in man. Although there is some indication that the sensitivity of the adrenal gland to stimulation may vary during the course of the day, the blood level of ACTH itself is known to fluctuate rhythmically. The peak output of corticosteroids in the rat occurs at about 4 p.m., at a time when the pituitary ACTH content is lowest in males, but at its highest in females. Damage to the brain which produces coma, interferes with the diurnal rhythm, as do suprasellar lesions in the temporal lobe and midbrain. The processes involved in the circadian rhythm of ACTH secretion are further discussed on p. 136.

Many observations attest to the fact that a balance exists between the blood concentration of adrenal hormones and the secretion of ACTH (Figure 7.1). Accordingly, chronic administration of adrenal corticoids inhibits pituitary secretion and leads to atrophy of the adrenal cortex. Removal of one adrenal gland lowers the blood concentration of corticoids and induces compensatory hypertrophy of the remaining organ, an effect which can be prevented by hypophysectomy or the administration of adrenal hormone. Increased blood levels of ACTH after complete adrenalectomy can now be measured, and fall after the injection of corticosteroids. By themselves the observations outlined above do not account for the increased output of ACTH provoked by stressful stimuli; for the increased discharge, and consequent elevation of adrenal steroid secretion, must be superimposed upon normal resting levels unless, as was suggested by Sayers and Sayers (1947), stress increased the rate of utilization of adrenal hormones by the body and so lowered the 'resting' level. The latter postulate was subsequently invalidated when technical advances made it possible to measure the blood levels of adrenal hormones and it was found that stress always caused an abrupt rise, not fall, in corticoid concentration. Further, the discovery that noxious stimuli applied to adrenalectomized rats could increase the already high blood level of ACTH, underlined the inadequacies of a simple feedback concept, which had earlier become apparent with the discovery by de Groot and Harris (1950) that electrical stimulation of the floor of the third ventricle of the brain in the rabbit raised the output of ACTH, as indicated by the occurrence of lymphopenia, and that electrolytic lesions placed in the same area prevented the response to psychic stress.

The Brain and ACTH Secretion

Electrical stimulation of the hypothalamus is now known to enhance the secretion of ACTH in the rat, rabbit, guinea-pig, cat, dog, and monkey. Unfortunately, there is little consistency in the localization of the sensitive area, so that in the rabbit excitation in the region of the mamillary bodies elicited lymphopenia, while in the dog stimulation of the anterior hypothalamus or of the anterior median eminence has induced a fall in the white cell count due to an increased secretion of adrenal steroids. This widespread response may be due to the existence of a relatively diffuse field within the median eminence concerned with the regulation of ACTH activity, or may indicate that in some areas important afferent pathways are being affected. The level of adrenal function is increased by stimulation of regions of the brain other than the

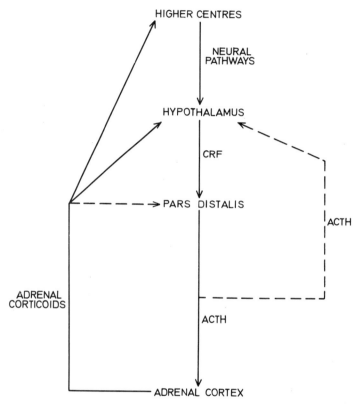

FIGURE 7.1 Schematic outline of the factors concerned in the control of ACTH secretion.

hypothalamus. Stimulation of the reticular formation in the midbrain is effective in this way, as might be expected from the role of this part of the brain in maintaining alertness, although the response observed depends somewhat upon the prevailing level of adrenal function. When this is low, facilitation of ACTH release generally occurs, but when the output of adrenal hormone is already high no effect, or some inhibition, may be apparent (Taylor, 1969). Excitation of limbic structures such as the amygdaloid nuclei, the orbital surface of the frontal lobe and the cingulate gyrus causes ACTH secretion in dogs, rabbits, and monkeys. On occasion, the secretion of ACTH is depressed, so indicating that there are neural mechanisms within the brain that inhibit the release of this hormone. The first hint came from work which showed that stimulation near the hippocampus abolished the adrenal response to bodily stress in cats and this observation was confirmed by the finding that stimulation of the hippocampus prevented the rise in blood adrenal hormone level that follows stimulation of the infundibulum. A reduction in the output of adrenal hormones has since been seen to follow stimulation of the septum, hippocampus, and even of the anterior hypothalamus, in several species including man. There is also evidence that midbrain areas are concerned in the inhibition of ACTH secretion (Mangili, Motta, and Martini, 1966).

In general, the effects of damage to the hypothalamus and other areas are the converse of those brought about by electrical stimulation. Lesions in the tuberal and median eminence region block stress-induced increases in ACTH secretion in all mammalian species tested, but do not always cause adrenal atrophy. From the fact that several lesions in the basal hypothalamus are more effective in reducing the ACTH response to stress than any single lesion, Brodish (1963) concluded that the entire region of the ventral hypothalamus, from the optic chiasma to the mamillary bodies, is involved in the control of ACTH secretion. Bilateral destruction of the amygdaloid nuclei, or of the afferent pathways to the hypothalamus, sometimes depresses pituitary-adrenocortical activity, but this is not a consistent finding and the damage does not routinely cause adrenal atrophy. Lesions in the lateral septal area or anterior cingulate gyrus have blocked the stress-invoked release of adrenal corticosterone in the rat (Usher, Kasper, and Birmingham, 1967), and transection of the brain stem at a midbrain level can abolish the stress-induced elevation in adrenal activity, although the resting secretion of adrenal hormones may be increased. This has led to the suggestion that the prevailing tone of midbrain activity may tend toward inhibition of ACTH release, and, through the reticular formation, be implicated in the diurnal rhythm in adrenal function. Support for this view has come from work with hypothalamic islands, for in rats studied four weeks after this operation the plasma concentration of corticosterone was higher than that of control animals, and the diurnal rhythm of ACTH secretion was lost (Halász, Slusher, and Gorski, 1967). Ether anaesthesia and unilateral adrenalectomy increased the blood level of steroid, and when animals with incomplete islands were studied it became clear that the anterior connections of the hypothalamus were important, for disruption of these alone blocked the circadian rhythm. On the other hand, while isolation of the hypothalamus prevented the normal adrenal response to audiogenic stress, Feldman, Conforti, Chowers, and Davidson (1968), considered that the resting level of plasma corticosterone was not elevated.

Removal of almost all of the brain is a drastic procedure in the investigation of the factors concerned in the control of ACTH secretion, but the results obtained have aroused much interest. In the studies of Egdahl and his colleagues on dogs (Egdahl, 1968) the removal of all the brain above the hypothalamus was compatible with an elevated secretion of adrenal hormones. This was, perhaps, not surprising, but the further removal of the hypothalamus did not depress adrenal function, in contrast to the situation after transection of the spinal cord in the neck when adrenal steroid output was low. These observations were accounted for on the basis of the release from the hind brain of a humoral factor which passed into the systemic circulation to reach the pituitary gland and so stimulate the secretion of ACTH. Under normal circumstances the release of the hind brain factor was believed to be chronically inhibited by some structure above the midbrain which was removed in the course of the drastic ablation. However, later experiments on dogs established that the secretion of adrenal corticoids remained high after removal of the entire brain, or of the brain and spinal cord, so that the operation of a hind brain factor in this response is unlikely. In the rat, serial ablation of the brain rostral to the superior colliculi, except for the hypothalamus, does not produce a sustained increase in adrenal function, but 'deafferentation' of the hypothalamus (p. 41) brings about an increase in adrenal weight and histological signs of a hypersecretion of ACTH. The response in the dog has been attributed to the action on the pituitary of substances released from tissue damaged in the course of surgery, or to the chronic leakage of ACTH from relatively ischaemic and abnormally permeable pituitary cells, but Egdahl (1968) has shown that the isolated pituitary gland remains well vascularized and points to the work of others on cats which have survived for a year after brain removal down to the pituitary gland and with no hypothalamic remnant. The adrenals of these animals were of normal size at autopsy and secreted sufficient corticoids to cope with the stress of infection. It may, or may not, be significant that extracts of cerebral cortex of unknown constitution, but possibly polypeptide in nature, prevent the afternoon rise in corticosteroid secretion in the rat. It has also been suggested that an increased sympathetic drive is important for the adrenal hyperactivity, since removal of the superior cervical sympathetic ganglia exerted a depressive effect. Some general significance of this work is indicated by recent observations of the continued secretion of adrenal steroids in substantial quantity after isolation of the pituitary gland of the monkey (Kendall and Roth, 1969). Surprisingly, the plasma level of 11-hydroxycorticosteroids was maintained at the pre-operative level after pituitary stalk section.

Although it is clear that the brain plays a dominant role in the control of ACTH secretion, the feedback action of adrenal corticoids must not be disregarded. The inhibitory effects of high blood levels of adrenal hormones upon ACTH release may vary in intensity from species to species. In rats, adrenal steroids block the response to a variety of stresses and there are indications of a linear relationship between the amount of hormone given and the inhibition observed. In men and dogs, rather small doses of corticoids suffice to suppress the resting secretion of ACTH, but it is much more difficult to prevent a discharge induced by stress. For example, cortisone administration fails to block the pituitary-adrenal response to laparotomy in patients, even when given in amounts which produce a plasma concentration which is far greater than that obtainable by release of endogenous hormone. Stressful stimuli, such as exposure to ether or surgery, elicit the discharge of ACTH from the pituitary gland very rapidly, with increases in the blood level of the hormone being detected within 150 seconds of exposure. The adrenocorticotrophic response to stress in adrenalectomized rats is markedly increased above that of normal animals, perhaps because endogenous hormones, among other factors, control the excitability of the hypophysis to the corticotrophin releasing factor. On the other hand, depression of the circulating adrenal hormone concentration by partial or complete adrenalectomy acts much more slowly and a maintained rise in ACTH secretion may not be apparent for many hours. Here, of course, it is necessary to disregard the initial rise in ACTH secretion induced by the surgery.

Lesions in the median eminence block compensatory adrenal hypertrophy and the increased ACTH secretion which normally occurs after bilateral adrenalectomy. Taken alone these observations imply that the feedback action of adrenal hormones is exerted upon the brain, but the situation is somewhat more complex. Since high doses of cortisol can cause adrenal atrophy in rats and dogs in which the median eminence has been destroyed, and the infusion of cortisol into the sella turcica causes adrenal atrophy, evidence is also available for an action of the steroid directly upon the pituitary gland. Local application of cortisone in agar to the tuberal region of the hypothalamus in cats and rats inhibits the secretion of adrenal steroids, while implants of cortisol into the median eminence and anteromedial hypothalamus have abolished compensatory hypertrophy in rats and depressed adrenal function on several occasions. Similar implants into the pituitary gland have proved ineffective (Fortier, 1966; Mangili, Motta, and Martini, 1966). Such experiments involving the implantation of solid hormone are open to the criticism that the median eminence with its primary plexus of portal vessels provides an ideal site from which diffusion of hormone to the pituitary gland can occur, but the high sensitivity of the pituitary gland to hormone which might be expected on this basis has not been demonstrated. Corticosterone implanted into the hypothalamus of rats was most effective in suppressing ACTH secretion when located between the optic chiasma and median eminence, and not when closest to the portal vessels. Indeed, the inhibitory action of the steroid decreased steadily as the pituitary stalk was approached (Smelik, 1969). Labelled cortisol appears to be concentrated selectively in the infundibular region of the hypothalamus, and not in the pars distalis. Further, the study of the electrical activity of single neurones in the hypothalamus and midbrain showed that units sensitive to dexamethasone (in that the firing rate was altered, and usually depressed) were scattered over wide areas. Steroid sensitive cells were not found in the cortex, dorsal hippocampus, or thalamus (Steiner, Ruf, and Akert, 1969). Overall, it seems that the case for a direct action of adrenal hormones on the pituitary gland under physiological conditions is not convincing, although it has been found that the response of the pituitary gland to exogenous corticotrophin releasing factor (p. 72) can be inhibited by corticosteroids, and that the action of endogenous corticotrophin releasing factor on the gland was inhibited by the direct intrapituitary injection of corticosteroid (Yates, Brennan, and Urquhart, 1969). Possibly the most direct experiments in this connection have involved the injection of dexamethasone into rats in which the forebrain was removed, leaving the pituitary gland isolated in the floor of a largely empty skull. The synthetic steroid still suppressed ACTH secretion sufficiently to produce a fall in plasma corticosterone, though not to the extent which followed hypophysectomy (Dunn and Critchlow, 1969).

The feedback action of corticosteroid hormones upon the neural mechanisms controlling ACTH release is not confined to the hypothalamus, but also involves the hippocampus and amygdala. Cortisone implanted into the amygdala of the rat reduces corticosterone secretion on the part of the adrenal gland, both under basal and stressful conditions, whereas implants made into the hippocampus enhance the response of the adrenal gland to stress. Compensatory hypertrophy of the remaining adrenal gland after removal of its partner has been depressed by lesions in the midbrain, and changes in the release of ACTH have followed the local application of corticosteroid to the septal area and rostral midbrain (Mess and Martini, 1968). Here it seems necessary to distinguish between effects on the resting secretion of ACTH and the enhanced activity normally expected with stress. When minute amounts of cortisol were inplanted bilaterally into various brain structures in the rat it was found that implants made into the median eminence, amygdala, medial thalamus, rostral midbrain reticular formation, and basal septal area, suppressed ACTH release upon exposure to the stress of environmental change, whereas implants located in the ventral hippocampus increased the output of ACTH. On the other hand, the resting secretion of ACTH, which fell with implants in the median eminence, remained unaffected after the application of cortisol to the amygdala or rostral midbrain reticular formation (Bohus, Nyakas, and Lissák, 1968). The need for caution in evaluating the effects of implantation of steroids in the brain upon the release of pituitary hormones is underlined by the recent work of Kendall, Grimm and Shimshak (1969), who injected an emulsion of cortisol acetate or implanted pellets of dexamethasone into the brains of rats and determined whether the ether-induced release of ACTH was suppressed. They associated the effectiveness of the suppression of ACTH release with the possibility of ready access of the steroids to the ventricular system of the brain in that steroids could be carried in the cerebrospinal fluid in a ventrocaudal direction to reach the basal portion of the third ventricle to act on glucocorticoid feedback receptors in the basal hypothalamus. In this view, the regulatory action of steroids in controlling pituitary function would be exercised solely through the hypothalamus with little involvement of extrahypothalamic structures.

A feedback action of adrenocorticotrophic hormone itself upon the hypothalamus has been suggested from time to time on the basis of indirect evidence, such as the fact that treatment of adrenalectomized animals with exogenous ACTH has occasionally reduced pituitary weight and blocked the fall in pituitary trophic hormone content usually induced by stress. But this possibility was not taken seriously until it was found that the implantation of ACTH into the median eminence of intact or adrenalectomized rats depressed endogenous ACTH release, and that exogenous ACTH can block the discharge of this hormone from the pituitary gland. The increased concentration of the hypothalamic releasing factor for ACTH which normally occurs after adrenalectomy, is not evident after giving ACTH, so substantiating a possible action of the peptide upon the brain. Treatment of adrenalectomized rats with ACTH alters the size of the nuclei in certain hypothalamic cells, and there are indications that ACTH can act on the brain to alter the electroencephalogram and multiple unit activity in the hypothalamus (Sawyer, Kawakami, Meyerson, Whitmoyer, and Lilley, 1968). Synthetic ACTH delivered microelectrophoretically to hypothalamic neurones increased the rate of discharge before exercising an inhibition, so that there is little doubt that ACTH can act on the brain. Other evidence to this effect is cited in chapter 16, and the existence of this phenomenon, variously termed the 'short', 'direct', or 'auto' feedback mechanism, is now well founded (Mess and Martini, 1968; Martini, Fraschini, and Motta, 1968; Motta, Fraschini, and Martini, 1969), but its physiological significance is not clear.

The range and variety of the factors that interact in the control of ACTH secretion have led to attempts to apply control theory to this problem. Before this can be achieved it is necessary to resolve the system under study into a number of components to which appropriate scale values can be given. Yates, Brennan, and Urquhart (1969) decomposed the overall adrenoglucocorticoid system into the following processes and components: the adrenal cortices, the binding of cortisol by plasma proteins, the distribution of cortisol through-out extracellular fluids, the appropriate cortisol feedback points and character-istics, the nature of the effect in that the signal applied can only be positive or negative in character, and the distribution and metabolism of ACTH. The interaction of these factors is set out in the block diagram of the overall model system illustrated in Figure 7.2, which has been used in the simulation of physiological experiments on a computer. In the light of both computer and physiological observations the model is being progressively improved.

Corticotrophin Releasing Factor

Study of the corticotrophin releasing factor (CRF) initially proved difficult because of the readiness with which almost any strange or unusual procedure elicits ACTH secretion in an experimental animal. So many drugs and tissue extracts can enhance ACTH secretion that extreme caution is necessary before it can be concluded that an active agent is the true releasing factor. Indeed, it is now possible to draw up a list of criteria that must be met before any presumptive releasing factor can be unequivocably identified with the physiological material. These may be summarized as follows:

1. The substance must be detectable in the median eminence of the hypo-thalamus, where it is stored at the nerve endings for release into the primary plexus of the portal vessels.
2. The substance must be demonstrable in portal vessel blood under suitable experimental conditions and its concentration should be related to the level of secretion of the appropriate pituitary trophic hormone. The concentration in the portal vessels should be higher than that in systemic blood.
3. The substance should release, or inhibit the release of, a single pituitary hormone, for if the release of several hormones were affected simul-taneously the agent might be toxic, or act as a general metabolic stimulant, and so lack specificity.
4. The agent should act on pituitary tissue *in vitro*, or on pituitary tissue separated from the hypothalamus by division of the pituitary stalk or by hypothalamic lesions.
5. Since the substance must act on pituitary tissue it should not influence endocrine function after hypophysectomy. If target organ function does change with administration of releasing factor in hypophysectomized animals then contamination of the extract with trophic hormone is com-monly, but not always, indicated.
6. If a releasing factor is stored in the hypothalamus then it might be expected that its concentration should change under experimental conditions which profoundly influence pituitary function, such as castra-tion, adrenalectomy, thyroidectomy, or target organ hormone administra-tion.

From the point of view of the control of adrenocorticotrophin output, several of these criteria can be satisfied at present. Extracts of the median eminence cause the release of ACTH, although other substances can act likewise. Portal vessel blood collected from dogs has been shown to stimulate ACTH secre-tion (Porter and Jones, 1956), and, in the experiments of many, hypothalamic extracts which act on the pituitary to promote adrenal function do not affect

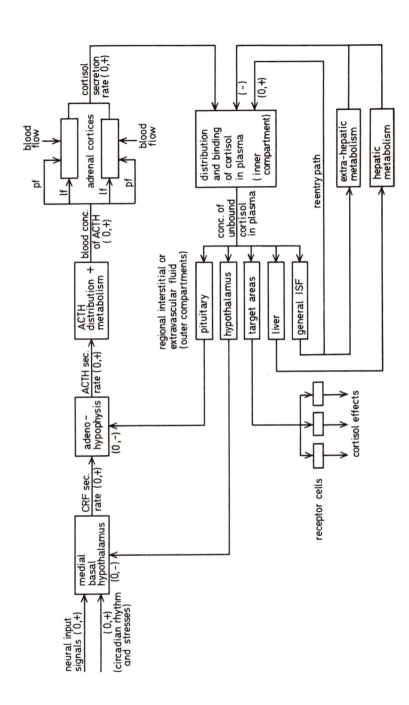

FIGURE 7.2 Model of the adrenal glucocorticoid control system. The term *lf* is indicative of the immediate effect of ACTH upon the adrenal gland (input forcing), while the pathways *pf* refer to the hypertrophic action of ACTH in the longer term (parametric forcing). From Yates, Brennan, and Urquhart (1969).

73

the release of other trophic hormones. Corticotrophin releasing activity has been demonstrated in the peripheral blood of rats after hypophysectomy, but did not appear in the blood of hypophysectomized animals in which the median eminence had also been destroyed, so indicating that the material was derived from the hypothalamus. Assays of CRF activity have successfully employed pituitary tissue incubated *in vitro*, while some of the changes in CRF secretion in various physiological states are outlined later on.

In the search for the factor controlling the release of ACTH, vasopressin (the antidiuretic hormone) has long been a prime candidate (Martini, 1966; McCann and Porter, 1969). It could be shown very easily that the administration of small doses of vasopressin caused ACTH secretion, and that vasopressin elicited release of the anterior pituitary hormone in rats in which the endogenous mechanisms controlling discharge had been disrupted by pituitary transplantation, lesions placed in the median eminence, treatment with a barbiturate and morphine, or massive doses of steroids. Further, stresses which caused the discharge of adrenocorticotrophic hormone also evoked vasopressin secretion and an antidiuresis, while ACTH release was depressed in animals in which diabetes insipidus had been produced by hypothalamic lesions. Vasopressin was also found to release ACTH from rat pituitary tissue cultivated *in vitro,* but the releasing activity could be separated from vasopressin by chromatography.

Secretion of vasopressin and of ACTH do not always go hand in hand. The correlation between ACTH secretion and antidiuresis is not absolute, in that production of the anterior pituitary hormone can be depressed by hypothalamic lesions in animals that do not have diabetes insipidus. Conversely, ACTH release continues in rats suffering from hereditary diabetes insipidus, that appear not to secrete vasopressin, although the increase in plasma corticosterone after stressful stimulation is less than that of their normal littermate controls.

Clinically, nicotine administration (which elicits release of antidiuretic hormone) continues to enhance ACTH secretion in patients with congenital diabetes insipidus. The secretion of vasopressin is blocked by ethanol, but such blockade does not depress ACTH secretion. The effects of vasopressin on ACTH discharge in the rat can be blocked by morphine, but such treatment is ineffective against extracts of hypothalamus. Frontal lobe stimulation in monkeys can increase the output of adrenal steroids without releasing the antidiuretic hormone. Evidence is also available for an extra-hypophysial action of vasopressin, for this hormone causes ascorbic acid depletion in hypophysectomized rats and can release steroids from the adrenal gland of the hypophysectomized dog.

Because of the close anatomical association between the adrenal cortex and medulla, and the fact that the stresses which increase the activity of one division of the adrenal gland often simultaneously affect the other, the possibility that adrenaline might act as a neurohumor has been considered and discarded. The injection of adrenaline, as well as of many other substances, calls forth ACTH secretion in the rat, but in the dog and man, adrenaline does not provoke the release of this trophic hormone. Stimulation of the dorsomedial hypothalamus in dogs increases adrenaline secretion without altering the output of adrenal steroids, and in situations (such as hypoglycaemia) where both adrenal responses are observed, that of the cortex precedes excitation of the medulla. Further, removal of the medulla does not disturb the customary

74

cortical response to stress, and the more radical procedure of complete surgical or pharmacological sympathectomy is likewise ineffective. It is, of course, quite possible that an adrenergic substance effective in promoting ACTH secretion may be released from the hypothalamus and not the adrenal medulla, for the hypothalamus is known to store sympathomimetic amines. However, adrenergic blocking agents given over a period long enough to permit adaptation, so that ACTH release no longer follows administration, do not inhibit the pituitary response to a range of stressful stimuli, though they do not prevent a response to adrenaline.

Corticotrophin release was the first property demonstrable by extracts of hypothalamus under proper test conditions but the effective substance, or substances, has not yet been obtained in pure form. Two types of agents with releasing properties (α- and β-corticotrophin releasing factor, α- and β-CRF) can be extracted from neurohypophysial tissue. α-CRF is related to α-melanocyte stimulating hormone (MSH) which lacks cystine and has properties suggestive of a role as a precursor of ACTH, while β-CRF is more closely related to vasopressin and is also present in the hypothalamus. β-CRF possesses a releasing activity some 50–100 times greater than α-CRF. Both substances are basic peptides, are unstable and possess inherent ACTH and MSH activity, probably through the occurrence of common sequences of amino acids in the molecule, and it is thus not surprising that vasopressin can, on occasion, mimic the effect of CRF. When CRF, lysine-vasopressin, and α-MSH (which also has CRF-like properties) were given for long periods to animals carrying grafted pituitary tissue, only CRF maintained the synthesis and release of ACTH by the transplants. Corticotrophin releasing factor is assayed by measurement of the amount of ACTH released from anterior pituitary tissue cultivated *in vitro,* or from the pituitary glands of rats treated with central depressant drugs or adrenocortical hormones to suppress endogenous ACTH secretion; one microgram is effective in this test. Adrenalectomy is known to treble the CRF content of the median eminence in rats killed 3 weeks after the operation, while treatment with cortisone reduces the CRF content of this area. So far as can be judged by present methods, CRF is not detectable in the peripheral blood of normal, unstressed, rats but appears after the stress of ether anaesthesia or haemorrhage. Hypophysectomy causes the appearance of CRF activity in the blood 2 or 3 days post-operatively, but this response can be suppressed by treatment with massive doses of corticosteroid. In this connection it has been shown that hypophysectomized rats carrying five pituitary glands derived from neonatal animals under the kidney capsule will respond to a stressful stimulus with a rise in plasma corticosterone. This is probably also due to the action upon the grafts of corticotrophin releasing factor (Purves and Sirett, 1967).

Control of Aldosterone Secretion

As noted earlier, in general terms the adrenal cortex may be regarded as secreting two kinds of steroids: the glucocorticoids concerned with the control of protein, carbohydrate and fat metabolism, and the mineralocorticoids, which are involved in the regulation of sodium and potassium excretion (Ashmore and Morgan, 1967; Grant, 1967; Mulrow, 1967). Aldosterone is the major hormone stimulating the reabsorption of sodium and the excretion of potassium and hydrogen, ammonium and magnesium ions by the kidney

(Mulrow, 1967); and although there is no doubt that the release of the gluco-corticoids is indirectly controlled by the brain through the release of ACTH, there has been much less agreement over the way in which the output of mineralocorticoids is governed.

Numerous attempts have been made to see whether the production of aldosterone falls when the adrenal is deprived of ACTH. By and large, removal of the pituitary gland depresses the secretion of aldosterone but the effect can be obscured by the stress of the operation needed to cannulate the adrenal vein prior to the collection of blood samples, and by some concurrent haemor-rhage. Both factors are known to increase aldosterone secretion and can be avoided by working with trained, conscious dogs. Following the administration of ACTH the output of both 17-hydroxycorticoids and aldosterone increases, while there is also evidence that treatment of dogs with glucocorticoids decreases the secretion of aldosterone, probably by limiting the release of ACTH from the pituitary gland. In these reactions adrenocorticotrophic hor-mone may affect aldosterone secretion through its general stimulatory action on steroidogenesis. Since the discovery of aldosterone many attempts have been made to determine whether the release of this hormone lies under neural control. These have often involved the placement of lesions in the brain, and a decrease in aldosterone has been recorded on a number of occasions. However, the output of other adrenal hormones was also depressed, so that the effect no doubt stemmed from a drop in ACTH secretion (Davis, 1967). Anxiety, apprehension, anger, exercise, and trauma increase aldosterone production, as well as that of ACTH (Gross, 1968).

The output of aldosterone is enhanced by conditions of chronic sodium depletion and is matched by an increased activity of the renin-angiotensin system of the kidney in man and dog (Figure 7.3). Plasma renin has been shown to rise during salt deprivation and fall during salt loading in normal human subjects, and corresponding changes in the plasma level of angio-tensin II also occur. After removing the kidneys of sodium-depleted hypo-physectomized dogs a marked fall in aldosterone secretion ensues. Similar observations have been made in sheep but, as Davis (1967) points out, the effects of nephrectomy on aldosterone secretion in the rat require further study in view of discrepant findings. Increases in potassium intake, or the administration of solutions of potassium chloride intravenously, have the opposite effect on aldosterone secretion to that of sodium, for mineralocorticoid output is increased. This response could be due to a direct effect on the adrenal cortex, or as a result of the loss of sodium that secondarily occurs in man. But fluctuations in potassium metabolism appear of minor significance in regulating aldosterone release.

Aldosterone secretion is influenced by humorally transmitted stimuli, as has been shown in work on animals with transplanted adrenals, and cross-circulation experiments in which blood from donor dogs with hyperaldoster-onism secondary to chronic thoracic inferior vena cava constriction was perfused through the isolated adrenals of normal recipient animals. Aldo-sterone secretion by the isolated adrenals increased markedly during cross-circulation and fell during control periods in the absence of changes in plasma sodium or potassium (Yankopoulos, Davis, Kliman, and Peterson, 1959). Haemorrhage also promotes aldosterone secretion, and as this procedure remains effective in hypophysectomized dogs, the pituitary can be eliminated as the source of the active factor. Further, experiments involving the removal

76

of the head, liver, or kidneys showed that only nephrectomy blocked the release of sodium-retaining material, and since the injection of saline extracts of kidneys greatly increased aldosterone production there is good evidence that the releasing agent comes from the kidneys (Davis, Carpenter, Ayers, Holman, and Bahn, 1961). In fact, it has come to be recognized that the kidney factor can be equated with the renin-angiotensin system of the kidneys. Renin is formed in the juxtaglomerular apparatus of each nephron, which is comprised of granular cells in the media of the afferent arteriole, the specialized tubular area at the junction of the ascending loop of Henle and the distal tubule, known as the macula densa, and mesangial cells in contact with both

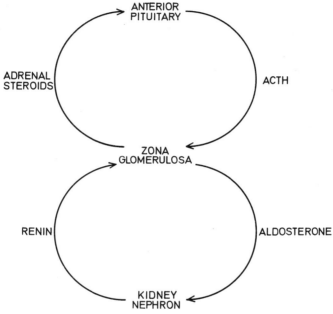

FIGURE 7.3 The control of the secretion of aldosterone by the zona glomerulosa of the adrenal cortex.

the granular cells and the macula densa. Renin is secreted into the blood and lymph, where angiotensin I is formed. Angiotensin I is a decapeptide which is then enzymically hydrolyzed to angiotensin II, an octapeptide, which stimulates the production of adrenal steroids (Figure 7.4). If the output of adrenal steroids is of sufficiently high magnitude, ACTH secretion will, in turn, be depressed to cause a fall in the output of aldosterone (Davis, 1967; Vander, 1967; Brown, Fraser, Lever, and Robertson, 1968). This concept is supported by a wide span of evidence, ranging from the fact that the administration of renin, angiotensin II, or kidney extracts increase aldosterone secretion, with lesser effects on glucocorticoid output, through studies of the effects of angiotensin II upon steroidogenesis, blockade of the response to renin by antibody formation, and measurement of the increased production of renin in cases of secondary hyperaldosteronism. This is not to say that the problem is solved, for the information available for the rat is not as clear as that for the dog and much of the evidence is inferential. It is remarkable, for example, that the administration of subpressor amounts of angiotensin can suppress the release of renin. Larger amounts of angiotensin, having a slight pressor effect, have been infused in man for many days and have maintained an increased secretion of aldosterone in the absence of a rise in cortisol production.

There is little certainty about the nature of the signal that evokes renin, and subsequently, aldosterone release. Some of the possibilities have been surveyed by Vander (1967), who considered that stimuli reaching the kidneys through the sympathetic nervous system could alter renin synthesis and release in at least three ways: by direct control of the macula densa cells, by inducing afferent arteriolar constriction, and, less directly, by reducing the glomerular filtration rate through constriction of the afferent arterioles. Since the renin-angiotensin system is also effective in maintaining the blood pressure, the pressor effects of sympathetic discharge become reinforced. It has long been known that constriction, but not closure, of the renal artery produces hypertension and that the kidney is equipped with a baroreceptor

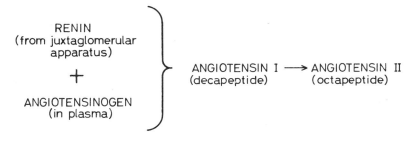

FIGURE 7.4 The renin-angiotensin system.

system, which could operate through the granular cells of the juxtaglomerular apparatus. Constriction of the aorta above the kidneys, which reduced the mean arterial pressure without depressing renal blood flow, has been found to cause the release of renin, while increased amounts of renin can be found in a kidney after constriction of the renal artery. The granulation and staining of the juxtaglomerular apparatus of that kidney becomes intensified, whereas converse changes occur in the control organ. Other anatomical and physiological observations indicate that renin release is controlled primarily by the macula densa, which could sense the osmolarity of the tubular fluid or its sodium concentration, although proof is needed. Vander (1967) suggests that renin release may vary inversely with the total load, rather than the concentration, of sodium delivered to the macula densa. Fluctuations in the sodium balance of an individual can also be detected indirectly through the changes in blood volume, renal perfusion pressure and renal blood flow that ensue. With renin secretion being influenced by these factors it is evident that a control mechanism for aldosterone is also provided. Further, the plasma concentration of aldosterone can be raised without increasing the production of the steroid, for in man the infusion of angiotensin may reduce hepatic blood flow by 25%, with a consequential depression in the metabolic clearance of aldosterone and other hormones.

Development of Neuroendocrine Control of Adrenal Function

In rats during the first few days after birth the response of the adrenal gland to a variety of stressors (surgery, temperature change, electric shock) is absent or depressed. Corresponding information is available for mice, dogs, rabbits, and guinea-pigs. This phenomenon is unexpected because it is known that adrenal function lies under pituitary control in the foetus and that the pituitary-adrenal system reacts appropriately to experimental intervention. Decapitation of the foetus (with loss of the pituitary gland) causes adrenal atrophy, and compensatory hypertrophy of the remaining gland occurs after unilateral adrenalectomy, while the injection of ACTH into the foetus stimulates adrenal growth and secretion. In the light of these facts it has been suggested that the non-responsiveness of the adrenal gland in the immediate post-natal period is due to the immaturity of the hypothalamic-pituitary system. However, rats, mice, rabbits, dogs, and guinea-pigs are born at widely different stages of development, with newborn guinea-pigs being sufficiently mature to adopt an independent existence almost at once. It has also become apparent that adrenal atrophy occurs in the rat foetus after removal of the brain and hypothalamus, while leaving the pituitary gland in place. This result implies that the brain controls pituitary and adrenal activity long before birth and is substantiated by the observation that the adrenal glands of encephalectomized (brainless) animals increased upon treatment with CRF, but only when the hypophysis was present (Jost, Dupouy, and Monchamp, 1966). Thus it is difficult to attribute the depressed reactivity of the adrenal gland seen after birth solely to developmental processes. On the other hand it is reasonable to suppose that the non-responsiveness reflects a relative degree of hyperfunction of the system during the first days of transition to extra-uterine life, with a consequent lack of functional reserve of ACTH secretion from the pituitary gland. Certainly the resting output of adrenal corticosteroids during this phase is much higher than that measured upon the appearance of responsiveness.

The Relationship between the Adrenal Medulla and Cortex

In mammals there is a close association between the adrenal cortex and adrenal medulla. Although the relationship is far from accidental, since embryonic tissue migration is necessary to bring the two components together, the reason for, or the benefits derived from, the partnership has not become apparent until recently. It has been known for some years that extra-adrenal chromaffin tissue, not surrounded by cortex, secretes only noradrenaline, whereas that of the adrenal gland can also produce adrenaline. The enzyme necessary for the methylation of noradrenaline to form adrenaline is phenylethanolamine-N-methyl-transferase (PNMT), present in the medulla. Surprisingly, the concentration of PNMT falls after hypophysectomy, and can be raised by ACTH, so that it appears that its activity is controlled to a significant degree by the glucocorticoids and is accordingly depressed in both experimental and clinical hypoadrenocorticism, when the production of adrenaline is known to be subnormal (Leach and Lipscomb, 1969). In this way, it seems, the adrenal cortex can influence the elaboration of an important medullary catecholamine; there are also indications that, in turn, the medulla can influence the production of steroids by the adrenal cortex.

8

Control of the Secretion
of Thyrotrophic Hormone

A substantial part of the relationship between the secretion of pituitary thyrotrophic hormone (TSH) and that of the thyroid hormones, as in the case of ACTH secretion and the adrenal cortex, falls in the realm of pure endocrinology. This is illustrated by the occurrence of pituitary hypertrophy in cases of hypothyroidism, and by the involution of the thyroid gland that follows hypophysectomy or the administration of thyroxine, which suppresses TSH secretion. Specific extracts of the pituitary gland cause hyperfunction of the thyroid gland upon injection, while interference with the synthesis of thyroid hormone by goitrogenic drugs (which free the pituitary gland from the brake of thyroid hormone) also induces hypertrophy of the thyroid gland. Indeed, with sensitive assay methods the relationship between the dose level of thyroid hormone (thyroxine) administered and the plasma TSH level can be seen to be quite close.

The actions of TSH are primarily confined to the control of the metabolism and hormone production of the thyroid gland, so that a detailed description of what the hormone does would largely duplicate a chapter on thyroid physiology (Bates and Condliffe, 1966). The hormone is a glycoprotein with a molecular weight of the order of 28 000, and has a half life of about 30 minutes in man and 15 minutes in the rat.

Study of the physiology of the thyroid gland, and of its neural control, received a great impetus with the advent of radioactive iodine for isotope work on the activity of this organ, for the release of this element from the thyroid gland is quantitatively related to the prevailing level of secretion of TSH. Isotopic procedures enable the level of thyroid function to be followed by measurement of the radioactivity in the neck region of intact individuals of suitable species for many days without the need for blood sampling, and the impetus given to research by this relatively simple method has not yet waned (Brown-Grant, 1966). Collaterally, bioassay techniques for the measurement of TSH in blood have been improved, and elevations of above normal can be reliably detected; it is less easy to quantitate subnormal levels of TSH in body fluids, although the great sensitivity of an immunoassay for TSH will resolve this problem. One early benefit derived from isotopic investigations was the confirmation of indications that the release of TSH could be affected very rapidly by suitable stimuli. Exposure of guinea-pigs to cold can, through the pituitary gland, produce detectable histological changes in the thyroid within 15 minutes and, in the rabbit, an increased output of labelled thyroid hormone can be observed in a few hours. Conversely, an injection of thyroid hormone suppresses TSH secretion within a matter of minutes.

The marked lability in the secretion of thyroid hormone alone might be taken to support the view that TSH secretion lies mostly under neural control, but this would be an oversimplification (Brown-Grant, 1966; Reichlin, 1966). While an important influence is exerted through the hypothalamus, the pituitary gland remains highly sensitive to the feedback action of thyroid hormone, even after the placement of massive lesions in the hypothalamus which depress the basal secretion of TSH. After isolation of the pituitary gland from the brain, the resting output of TSH, though lower than normal, is maintained at a much higher level than after hypophysectomy, but still can be depressed by thyroxine, though less of the thyroid hormone is required than in the intact animal. It is worth noting that the reduced basal level of TSH secretion does not necessarily prove that the release of this hormone is under the tonic control of the hypothalamus. The other endocrine and metabolic abnormalities that appear after disruption of the pituitary stalk could lead to a reduced thyroxine disposal rate, so that the system reaches equilibrium at a lower level of TSH secretion and hence of thyroid activity (Brown-Grant, 1967). However, experiments involving the insertion of fragments of thyroid tissue directly into the pars distalis indicate that secretion of TSH by the surrounding pituitary gland is suppressed, while comparison of the effect of local injection of microgram amounts of thyroxine into the hypothalamus with that of comparable placements in the pituitary gland of the rat, rabbit, and cat has shown that only injections made into the hypophysis readily inhibit the release of ^{131}I from the thyroid gland. Injections made into the hypothalamus do not elicit the same type of response, although inhibition of thyroid function has been described. The evidence in favour of a feedback action of thyroid hormone upon the hypothalamus has proved equivocal, for although the local application of thyroid hormone to the anterior hypothalamus has resulted in a depression of thyroid activity, it now appears that damage to this part of the brain may have been responsible, since a similar effect followed the implantation of muscle (Mess and Martini, 1968). The pituitary gland also reacts directly to *falls* in the concentration of thyroid hormone by increasing the output of TSH. This can be shown in hypophysectomized animals in which pituitary tissue is grafted to the anterior chamber of the eye, or in animals suffering from massive destruction of the hypothalamus, when the circulating level of thyroid hormone is decreased by hemithyroidectomy or by treatment with goitrogenic drugs.

The Environment and TSH Secretion

The existence of a connection between the emotional state of an individual and the level of thyroid function has long been taken for granted. Parry (1825) associated the onset of hyperthyroidism with emotional shock, as did Graves (1835) ten years later. Some 80 years ago, the Committee of the Clinical Society of London reported upon 109 patients suffering from hypothyroidism, and remarked that delusions and hallucinations occurred in nearly half of the cases, mainly where the disease was advanced. Insanity as a complication was noted in about the same proportion as delusions and hallucinations, and took the form of acute or chronic manias, dementia, or melancholia, with a marked predominance of suspicion and self-accusation. In such cases the mental state reflects the response of the brain to an abnormally low level of thyroid hormone; when excessive amounts are present, as in thyrotoxicosis, hyperexcitability is manifest. Thyrotoxicosis associated with exophthalmos is known as Graves' disease, although there is disagreement over the definition of this syndrome (McKenzie, 1968). On numerous occasions, an apparently sudden onset of this disease has been regarded as a consequence of a traumatic emotional episode such as a bereavement or separation and this relationship has been taken as evidence for the neural control of thyrotrophic hormone secretion, although the disease may already have been in progress and what was noted was an unusually spectacular personal response to the event. Curiously, in experimental animals emotional stress inhibits thyroid activity, yet after an exhaustive review of the literature dealing with the clinical effects of emotion on thyroid activity, Gibson (1962) concluded that there was no evidence that emotional stress produces hyperthyroidism in man. The effects of stress upon thyroid function in animals and man have been surveyed by Dewhurst, El Kabir, Harris, and Mandelbrote (1968). In animals, an inhibition of thyroid activity following an acute stress has been recorded in the majority of species studied, apart from some experiments in sheep where an increased secretion of thyroid hormone followed the emotional stimulus of a barking dog, or bangs, and lasted for up to 2 hours. In longer-term experiments on rats and rabbits, inhibition of thyroid function may persist for some days after a stress. With regard to man, the diversity of findings described in the literature made generalization difficult, but it seems that after the physical and emotional trauma of surgery an acute rise in the blood concentration of protein-bound iodine, which was often short-lived, occurred. Sometimes there was no change in the protein-bound iodine level, which provides an indirect measure of blood thyroid hormone concentration. More acute stresses, as applied through interviews, examinations, or films, have indicated that thyroid activation may occasionally ensue.

Despite the emphasis given to emotional factors in considerations of the pathogenesis of Graves' disease, recent work indicates that neural stimuli may be of little relevance in the induction of this condition. Alongside pituitary thyroid-stimulating hormone, the human thyroid gland can be influenced by a long-acting thyroid stimulator (commonly referred to as LATS), which is a gamma globulin acting as an antibody to some unknown constituent of the thyroid gland, although it is not clear how an antibody can stimulate the metabolism of this endocrine organ. The substance may be produced by the plasma cells and its secretion, unlike that of TSH by the pituitary gland, cannot be suppressed by treatment with thyroid hormone (McKenzie, 1968, 1969).

In view of the above information, what experimental indications are there for a neural control of TSH secretion? The strongest evidence comes from investigations involving electrical stimulation of the hypothalamus. Harris and Woods (1958) showed that in conscious unrestrained rabbits electrical stimulation of the anterior part of the median eminence close to the supraoptico-hypophysial tract accelerated the release of hormone from the thyroid gland, whereas stimulation of more posterior or superior areas of the hypothalamus, or of the pituitary gland, did not enhance thyroid activity. The effectiveness of the stimulation was increased following removal of the adrenal glands and maintenance of the animals on a constant dose of cortisone, for it became clear that the concomitant release of ACTH and of adrenal corticoids could suppress the response of the thyroid gland, probably because the TSH and ACTH fields of the hypothalamus overlap in the posterior region of the median eminence and simultaneous excitation of the adrenal cortex blocks TSH secretion. Electrical stimulation of the hypothalamus of cats and rabbits has since been shown to raise the plasma concentration of TSH some six-fold within 15–20 minutes, but excitation of other areas of the brain has proved much less effective (McCann and Porter, 1969).

Destruction of part of the hypothalamus by electrolytic lesions depresses thyrotrophic hormone secretion in all species studied, although the limits of the area shown to be influential in this way are not well defined. Lesions in the anterior basal hypothalamus reduce the compensatory hypertrophy of the thyroid tissue that normally follows removal of one lobe of the organ, and block the thyroid enlargement expected upon treatment with goitrogenic drugs. There is little doubt that the rostral hypothalamus is of great significance and that there need be no interference with the primary plexus of portal vessels. Damage to the portal vessels causes a greater depression of thyroid function than lesions of the hypothalamus alone. Complete deafferentation of the medial basal hypothalamus of rats moderately depressed thyrotrophic hormone secretion, but posterior frontal cuts in the hypothalamus, made approximately 1 mm behind the posterior margin of the optic chiasma, caused a marked decrease in thyroid activity (Halász, Florsheim, Corcorran, and Gorski, 1967). The latter cuts disconnected a region from the median eminence, extending from the level of the suprachiasmatic nuclei to the anterior border of the ventromedial nuclei, that was considered to be particularly concerned in the control of TSH secretion. Experiments involving the implantation of pituitary glands taken from neonatal rats into the hypothalamus of hypophysectomized adults, or recipients given goitrogen with or without thyroxine, have shown that thyroidectomy cells appear in the grafts located in the anterior hypothalamus between the suprachiasmatic and ventromedial nuclei and extending dorsally to the paraventricular nuclei (Flament-Durand and Desclin, 1968), and in an area more or less coincident with the zone delineated by Halász and his co-workers. From the studies involving electrical stimulation of the hypothalamus it would also seem that the region between the paraventricular nuclei and the infundibulum is particularly important, while lesions in this area impair thyroid function. However, thyroid activity can be normal after complete destruction of the supraoptic nuclei, the paraventricular nuclei, or other well demarcated hypothalamic structures. In large part, experimenters employing hypothalamic lesions have looked for an abolition of the thyroid hyperactivity that normally follows treatment with goitrogen. If this seemed to be suppressed then it was argued that the lesion had interfered with the release of TSH by the pituitary gland. Alongside disturbance of the pituitary response to a deficit of thyroid hormone, it has become apparent that the mean level of thyroid function is depressed, as shown by a decrease in thyroid weight and in the uptake and release of labelled iodine, and that the disturbance sets in within 6 hours of the operation. Despite the reduction in the basal secretion of TSH that occurs after hypothalamic damage the pituitary-thyroid axis can still respond to appropriate stimuli. Administration of goitrogenic drugs causes the output of TSH, though subnormal, to increase, whereas treatment with thyroxine inhibits TSH secretion. In fact, the sensitivity of the system in terms of the amount of thyroxine needed to block TSH secretion may be enhanced. Thus it is suggested that the tonic action of the hypothalamus sets the baseline level of TSH secretion and operates to reduce the sensitivity of the pituitary-thyroid axis to changes in the thyroid hormone level of the blood (Reichlin, 1966).

Rather little is known about the extra-hypothalamic control of TSH secretion. Although changes in thyroid function have been reported upon electrical stimulation of the limbic system, removal of the forebrain rostral to the superior colliculus has not depressed the goitrogenic effect of thiouracil, nor the enhancement of thyroid function seen on exposure to cold (van Beugen and van der Werff ten Bosch, 1961). The habenular nuclei, located posteriorly above the thalamus and beneath the hippocampus, are considered by some to be important in the regulation of TSH release. Lesions of these nuclei have depressed the goitrogenic response to thiouracil, while the secretion of TSH appears to be increased beyond the usual level in animals with habenular lesions and exposed to cold or subjected to hemithyroidectomy. In explanation, it has been suggested that any habenular control over TSH release may be exercised only under extreme conditions, when an individual is subjected to changes which may overburden the system (Mess and Martini, 1968).

Temperature and Thyroid Activity

There are many indications that exposure to cold brings about an increase in thyroid activity. These include increases in thyroid weight, histological changes in the thyroid and pituitary glands, increased turnover of radioactive iodine, and an increase in the plasma level of TSH. The rise in oxygen consumption and in metabolic rate that follows exposure to cold has been known for many years. As might be expected, with elevation of the environmental temperature there is a progressive depression in thyroid activity (Brown-Grant, 1966; Reichlin, 1966).

The mechanisms involved in the response of the thyroid gland to exposure of the individual to cold are closely related to those concerned with the maintenance of body temperature, for both are represented within the hypothalamus. Lesions placed in the anterior hypothalamus interfere with heat loss mechanisms, as well as with the release of TSH. Panting or skin vasodilatation does not occur upon exposure to a warm environment so that the individual (experimental animal or patient) develops a hyperthermia which can prove fatal. Electrical stimulation of the anterior hypothalamus initiates panting and skin vasodilatation, so that the body temperature falls. On the other hand, a region of the posterior hypothalamus seems to be concerned with the control of heat production, for lesions located dorsolaterally to the mamillary bodies eliminate the shivering and piloerection that normally follow exposure to cold and hypothermia ensues. Muscular tremors like those associated with shivering have been observed upon electrical stimulation of the posterior hypothalamus. Under physiological conditions, the two thermoregulatory centres in the hypothalamus act reciprocally to maintain body temperature at a normal level in response to information provided by thermosensitive receptors in the skin and diencephalon. The centrally located thermoreceptors are to be found in the preoptic area where some neurons discharge at an increased frequency as the adjacent hypothalamus is heated. Local heating of the anterior hypothalamus also induces sweating, panting, and vasodilatation, while cooling of this part of the brain by much less than one degree can cause a detectable fall in skin temperature.

The secretion of TSH is increased by cooling the hypothalamus. In the work of Andersson and his co-workers on goats (Andersson, 1964), for example, increased thyroid activity was demonstrable by a rise in plasma protein-bound radioactive iodine concentration and a fall in the level of radioactivity in the thyroid gland. The rise in plasma protein-bound iodine became apparent after 30 minutes, but the steepest rise generally occurred between 30 and 150 minutes after the onset of central cooling. If the preoptic or anterior hypothalamic region was cooled for less than 30 minutes the change in thyroid function set in after the hypothalamus had returned to its normal temperature. Destruction of the median eminence completely blocked the thyroidal response to cooling of the heat loss centre, and this response could also be blocked by the administration of thyroxine, though the hormone had to be given two hours before the test for best effect. Local warming of the anterior hypothalamus blocked the increase in thyroid activity that normally occurred when large amounts of ice-water were introduced into the rumen. Rather similar observations have been made in the rat (Reichlin, 1964) but are necessarily less precise because of the small size of the brain when compared with the thermodes employed.

Activation of the thyroid gland does not represent the first physiological line of defence against cold. Cutaneous vasoconstriction and shivering are much more effective and important. Indeed, exposure to severe cold acts as a stressful stimulus and may lead to inhibition of the release of thyroid hormone. On other occasions, falls in the blood level of thyroid hormone have been recorded upon exposure to cold, perhaps due to increased degradation in the periphery, and some activation of TSH secretion may be expected through the operation of a feedback mechanism. The neural mechanisms involved in the control of TSH secretion may not be able to sustain an increased output of this trophic hormone for long periods.

Exposure to cold can influence neuroendocrine function in other ways. In the goat, exposure to cold brings about an immediate increase in the release of sympathetic amines, and an increase in adrenaline excretion occurs upon cooling of the heat loss centre. When the animal was kept in a cold environment, and the heat loss centre warmed, the excretion of catecholamines remained low, but increased considerably once the local warming was stopped, particularly if pronounced hypothermia had developed in the course of the experiment. The sympathetic amines released in this way seem to be important in thermogenesis and in facilitating shivering (Andersson, 1964).

The responses outlined above may reflect different aspects of the action of a physiological thermostat located in the hypothalamus, so that the need for enhanced heat production upon exposure to cold would be reflected by adjustment of the mechanism to a higher setting so that processes for both the conservation and production of heat are set in train. In addition to the reactions involving vasomotor responses, shivering, activation of the sympathetic system, and TSH release, food intake is also raised. To cite the goat once more, a rise in hay consumption to a level 40% above that of the control has been observed during a seven-day period of cooling of the rostral ventromedial hypothalamus.

Stress and Thyroid Activity

Many noxious stimuli (restraint, immobilization, haemorrhage, surgery) inhibit the release of hormone from the thyroid gland. Such stimuli often elicit ACTH release and so enhance adrenocortical function, but the inhibitory effect on the thyroid gland is not secondary to the rise in corticosteroid output because it is not blocked by adrenalectomy. There are indications of the existence of an inverse relationship between the release of TSH and of ACTH, so that if the secretion of ACTH is suppressed the output of TSH can no longer be inhibited. On the other hand, electrical stimulation in the region of the mid and posterior median eminence in rabbits enhances thyroid function most readily after adrenalectomy, while thyroid activity is reduced by cortisone administration, so that some interaction between adrenal and thyroid function clearly occurs. It might be expected that cortisone would interfere with the secretion of TSH by an action on the hypothalamus, but this steroid could still inhibit the release of thyrotrophic hormone after the pituitary gland was isolated from the hypothalamus by section of the pituitary stalk (Brown-Grant, 1966; Reichlin, 1966).

Thyrotrophin Releasing Factor

Extracts of hypothalamic tissue will release TSH from transplanted pituitary tissue or, in much higher dosage, from the gland *in situ*, whereas extracts of cerebral cortex do not have this effect. The extract may be applied to grafts of pars distalis located in the anterior chamber of the eye or added to the fluid in which glands are cultured *in vitro*. If active, such extracts are considered to contain thyrotrophin releasing factor (TRF). Other techniques for the assay of TRF activity include measurement of the release of ^{131}I from the thyroid gland of rats, or measurement of the increase in the plasma level of TSH which also occurs. However, the assay of TSH presents difficulties of its own. It is always necessary to ensure that the material tested for TRF is free from TSH, for the pituitary hormone has been found in the hypothalamic pieces or stalk-median eminence tissue which provides the raw material for extraction. It is entirely possible that the releasing factors may be produced by neurones situated far from the median eminence, and transported to the primary plexus of portal vessels in a similar way to that employed by the histologically-visible neurosecretion. This suggestion has been tested by destroying various parts of the hypothalamus and measuring the content of the releasing factors in the median eminence to determine whether the con-centration is depressed, as might be expected after elimination of the cells synthesizing the materials. Suprachiasmatic, paraventricular, and arcuate-ventromedial lesions have all been shown to significantly reduce the content of TRF at the level of the median eminence and these results have been inter-preted as indicating that TRF is synthesized over a large hypothalamic area (Mess and Martini, 1968).

That a thyrotrophic hormone releasing factor is discharged into the hypo-physial portal vessels after electrical stimulation of the brain has been shown directly by the work of Averill, Salaman, and Worthington (1966), who collected portal vessel blood from rats before and after hypothalamic stimulation, and infused it into the pituitary gland of rabbits. In the absence of hypothalamic stimulation the secretion of TSH by the rabbit tended to fall with the infusion of rat portal vessel blood, whereas samples collected during excitation of the brain maintained the output of TSH or caused it to rise. Much could be learned from further experiments of a similar nature, but they are technically extremely difficult. The concentration of TRF in the hypothalamus of the rat is increased by thyroidectomy, but is not reduced upon the administration of thyroxine (Sinha and Meites, 1965). On the other hand, it seems that large doses of thyroid hormone can block the action of TRF in promoting the release of TSH from pituitary tissue under *in vivo* or *in vitro* conditions (Averill, 1969). There is disagreement over the question whether inhibitors of protein synthesis can block this effect (Mittler, Redding, and Schally, 1969).

The immense effort put into the separation and purification of TRF is reminiscent of the early heroic days of endocrinology. Then, two tons of ovaries were used to provide starting material for the isolation of oestrogen; now, a ton of hypothalamic tissue yielded less than 2 mg of a substance with high TRF activity—and this with modern physicochemical procedures. In the light of this fact it is perhaps to be expected that recent studies (Schally, Arimura, Bowers, Kastin, Sawano, and Redding, 1968; Schally, Redding, Bowers, and Barrett, 1969) have established that porcine TRF is active in nanogram amounts in causing TSH release in mice, and that the activity of 10 picograms (0·01 nanogram) can be detected with rat pituitary tissue *in vitro*. It has been calculated that TRF may release 200 to 2000 times its own weight of TSH.

Agreement has not yet been reached over the nature of TRF. It was once considered to be a polypeptide, but now it seems that the activity is not lost after treatment with trypsin, pepsin, pronase, carboxypeptidase, or amino-peptidase. According to one group of workers, (Bøler, Enzmann, Folkers, Bowers, and Schally, 1969) pig TRF contains three amino acids (histidine, proline, and glutamic acid) as L-pyroglutamyl-L-histidyl-L-proline amide, with the following structure:

$$
\begin{array}{c}
CH_2 \text{------} CH_2 \\
\end{array}
$$

The substance was synthesized and behaved as the natural product in several chromatographic and biological systems. Another group (Burgus, Dunn, Ward, Vale, Amoss, and Guillemin, 1969) reported that six synthetic isomeric tripeptides containing histidine, proline, and glutamic acid lacked biological TRF-like activity, but that after acetylation the derivatives of the sequence glutamic acid—histidine—proline readily elicited TSH release. Working with ovine material, these investigators (Burgus, Dunn, Desiderio, and Guillemin, 1969) have reached conclusions similar to those of Bøler et al. (1969) on the nature of TRF, although there are differences in matters of detail. Information of this kind is of great importance in elucidating the structure of the natural releasing factor, and indicates that ample material capable of promoting TSH release may soon become available for experimental work.

Although several investigators have suggested that the neurohypophysis is involved in the control of TSH secretion, no satisfactory evidence for the existence of such a relationship has been forthcoming. Disease or disturbance of the neurohypophysis does not alter thyroid activity in any consistent fashion, and such effects as do arise can be accounted for by the changes in water balance or food intake that ensue. Study of the effects on the thyroid gland of vasopressin and oxytocin has led to contradictory results which preclude generalization. Nevertheless, in the rabbit and dog it has been shown that vasopressin can act directly on the thyroid gland to release thyroxine. Melanocytestimulating hormone can also enhance thyroid activity without affecting the pituitary gland, and may be regarded as possessing a TSH-like activity which is possibly due to the occurrence of similar sequences of amino acids in melanocytestimulating hormone and TSH.

9

Control of the Secretion of Growth Hormone

Understanding of the physiology of growth hormone has been transformed in recent years by the development of sensitive immunological assay procedures. And because most work has been done with human growth hormone it is probably fair to add that more is known of the factors controlling growth hormone in man than in lower species. The remarkable sensitivity of the immunoassay for human growth hormone is illustrated by the observation that routine assays in a clinical laboratory employ plasma diluted some 20 times, in which the basal concentration of growth hormone may be no greater than 0·5 mµg/ml. While it has been known for many years that the pituitary gland secretes a growth-promoting principle, both the release and action of the factor was considered to be slow and prolonged, whereas it is now realized that the secretion rate is highly labile and that the half life of human growth hormone is of the order of 20–30 minutes; that of the rat is about 7·5 minutes.

Growth hormone (GH) or somatotrophin (STH) acts on every tissue in the body and possesses glycotropic, pancreotropic, glycostatic, diabetogenic, ketogenic, and lipotropic properties. Of importance, is the fact that the physiological effects of testosterone, oestrogens, and ACTH are also enhanced. Whereas growth hormone and prolactin are considered to be separate entities in many mammalian species, it has been extremely difficult to prove that both factors are secreted separately by the human pituitary gland. Highly purified human growth hormone shows mammary gland stimulating (mammogenic and lactogenic) activities in animals and man, and the ratio between the somatotrophic and prolactin-like actions varies considerably in different purified samples of human growth hormone. It is not yet clear whether the different ratios reflect the occurrence of two separate hormones in varying proportion, or a changing incidence of molecular groupings responsible for the effect. What is certain is that a human growth hormone preparation which showed lactogenic activity in rats, greatly improved lactation in women (Lyons, Li, Ahmad, and Rice-Wray, 1968).

An important action of growth hormone is, as its name suggests, to promote growth and it might be thus expected that higher plasma levels of this hormone may be found in children than in adults. High levels are found for a short period after birth, and are frequently recorded during adolescence, but recent information shows that the plasma level of growth hormone can fluctuate considerably in young and old, and it seems that only while fasting and at rest do marked differences between adults and children show up (Hunter, Rigal, and Sukkar, 1968).

The regulation of growth or somatotrophic hormone secretion has been reviewed recently by Reichlin (1966) and Pecile and Müller (1966), who well summarize knowledge prior to the application of immunoassay. Attempts to follow the secretion of growth hormone at different ages were largely based upon changes in the GH content of the pituitary gland, as well as in the percentage of the acidophil cells believed to produce the hormone. Since growth was disturbed by such procedures as pituitary stalk section, pituitary transplantation, and brain lesions, it appeared that the secretion of growth hormone was depressed, but the relationship of the retardation of growth to a paucity of GH was not beyond question. For growth hormone secretion is reduced in cases of hypothyroidism (which also ensues after interference with the pituitary stalk), as well as in cases of adrenal failure, and malfunction of these target organs could contribute to the slowing of body growth. Large amounts of adrenal corticoids act antagonistically to growth hormone. Injured or sick animals, or children, grow less well than normally so that it might be argued solely on this basis that stress inhibits somatotrophin secretion. However, skeletal maturation proceeds normally and a fall in food intake as well as in muscular activity must also be taken into consideration. As will appear, stress actually increases growth hormone secretion in man.

Three major influences are operative in the control of growth hormone secretion. These are neurogenic, often stressful, stimuli, the intracellular glucose concentration within the brain, and the blood level of amino acids (Figure 9.1).

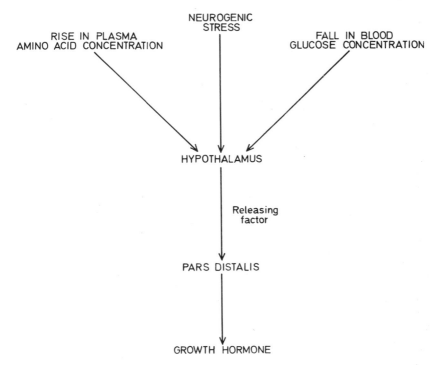

FIGURE 9.1 The factors concerned in the control of the secretion of growth hormone.

Stress and Growth Hormone Secretion

It is quite clear that under stressful circumstances in man the secretion of growth hormone is increased. Sudden loud noises, surprise, pain, fear, and other psychic stimuli are highly effective; the plasma level of GH has been shown to rise some forty times in the course of a non-traumatic dental examination. Growth hormone levels in the plasma have been found to be high in patients suffering injury and treated in a hospital casualty department. The levels fell with treatment. It has been shown in squirrel monkeys that transfer from the home cage to a restraining chair brought about a rise in circulating GH which declined in the course of an hour. Often the rise in plasma somatotrophin is paralleled by a rise in plasma cortisol under conditions of stress, but this relationship is not found when, for example, growth hormone output is elicited by the infusion of insulin. Conversely, the effect of stress in causing growth hormone secretion is not blocked by the administration of glucose, which normally depresses the output of GH. Stress is thus a potent factor in promoting the discharge of the hormone but there is evidence that the growth retardation seen in emotionally deprived children may be due to a psychically-induced *inhibition* of growth hormone secretion, for transfer from a difficult home environment to a protective institution has been followed by a marked acceleration in growth rate, which cannot be accounted for upon a dietary basis. Exercise also calls forth the release of growth hormone and may act through a concomitant fall in the blood glucose level. Systematic fondling of rats has been shown to enhance growth and the efficiency of food utilization, and it is pertinent that the plasma GH levels in rats that were 'gentled' by handling for 5–10 minutes each day over a 2-week period were significantly higher than the depressed GH levels in animals left unattended except for being fed and watered every third day (Schalch and Reichlin, 1968).

In contrast to the unanimity of the findings in primates, there is controversy over the effects of stress in rats, for some argue that stress elicits growth hormone release in this rodent, and others put forward evidence for an inhibition of the output of this factor. For example, Müller, Arimura, Sawano, Saito, and Schally (1967) found that stresses such as exposure to cold (4°C) caused a fall in pituitary GH content, and the appearance of growth hormone releasing factor in the plasma. The growth hormone was measured by bioassay in this work. On the other hand, Schalch and Reichlin (1968) report that the stress of forced swimming for 4 minutes, ether anaesthesia, and cardiac puncture, or exposure to cold reduced the plasma levels of growth hormone, as measured with an immunoassay developed especially for rat growth hormone. Cross-reaction with other pituitary hormones such as prolactin, gonadotrophins, TSH, and ACTH did not seem to occur so that it was suggested that current bioassays measure growth promoting materials other than growth hormone, such as prolactin. The classic bioassay for growth hormone is based upon its action in widening the tibial cartilage of rats, and parallel experiments employing both bioassay and immunoassay in the evaluation of the effect of insulin-induced hypoglycaemia, or of exposure to cold, on the pituitary content of growth hormone in rats, were carried out by Daughaday, Peake, Birge, and Mariz (1968). Curiously, the pituitary content was not depressed by either procedure as determined by bio- or immunoassay. No explanation for this apparent lack of response is currently available. Others (Müller and Pecile, 1968) have observed that exposure of rats to 42°C for 1 or 4 hours, forced swimming for 1 hour in a bath at 28°C, the subcutaneous injection of formalin, or exposure to cold for 5 minutes, did not alter pituitary growth hormone content, although exposure to cold for 60 minutes was effective. Thus not all stresses acted alike in causing GH secretion and the secretion of this hormone did not invariably parallel that of ACTH, which is evoked by formalin or vigorous exercise. It is clear that much remains to be learned about the effect of stress on somatotrophin release in the rat.

Glucose, Amino Acids, and Growth Hormone Secretion

There is a variety of evidence that indicates that the mechanism controlling GH secretion is responsive to the blood glucose level, or, more particularly, to the intracellular concentration of glucose. Hypoglycaemia produced by insulin is a potent stimulus, as may be a rapidly falling blood level of glucose which nevertheless remains above hypoglycaemic levels. Insulin itself does not seem able to alter GH secretion, since there is no increase in the output of the pituitary factor when glucose is infused together with insulin so that the blood glucose concentration remains constant. The intracellular utilization of glucose can be blocked by 2-deoxy-α-glucose, and when this substance is administered the blood sugar rises. Nevertheless, symptoms of hypoglycaemia appear and the plasma level of GH is elevated. Measurement of plasma growth hormone under conditions of fasting, or of continuous intravenous infusion of glucose, has shown that the blood level fluctuates somewhat rhythmically, rather than remaining at a constant level. Continuous infusion of glucose failed to suppress somatotrophin secretion and the peak levels at night were higher than those seen during the day (Glick and Goldsmith, 1968; Hunter, Rigal, and Sukkar, 1968).

The relationship of growth hormone secretion to the metabolism of protein is less clear. It is generally considered that a rise in plasma amino acids stimulates GH secretion so that in turn the amino acid level falls by the incorporation of them into protein. Eating meat, or the intravenous infusion of arginine, are potent stimuli for somatotrophin release in adult women, but men respond less consistently to the amino acid. Interestingly, the response of men to arginine was enhanced by treatment with stilboestrol (Rabinowitz, Merimee, Nelson, Schultz, and Burgess, 1968). Here it is of interest that women tend to have higher plasma GH levels than men, and that the mean values may be higher immediately after ovulation than before.

Neural Control of Growth Hormone Release

Until the advent of sensitive assays for growth hormone, information concerning the possibility of a neural control of the release of this pituitary factor was fragmentary and indicative only. The disturbances of growth that had been reported to occur after interruption of the pituitary stalk could be attributed to a depression of the release of ACTH and TSH and not solely to a deficiency of growth hormone; and the increase in the sensitivity of stalk-sectioned animals to insulin that had been described was equally difficult to interpret. Work with transplanted pituitary tissue has since shown that the secretion of GH by such grafts is very low, but increases when the pituitary tissue is replaced in contact with the median eminence. Deafferentation of the hypothalamus by surgical means in the rat retarded body growth, and reduced the plasma concentration and pituitary content of GH by rather more than 50%, as measured by immunoassay. Nevertheless, substantial amounts of growth hormone were still secreted. Since pituitary growth hormone levels were equally reduced when either the lateral, dorsal, or posterior connections of the hypophysiotrophic area were interrupted, and the slight retardation in body growth seen in animals with complete deafferentation was not evident in rats in which incomplete islands were prepared, it has been suggested that although neural afferents to the region are required for normal GH secretion, the afferents concerned are not specific in organization (Halász, 1968).

It is also difficult to interpret many of the early studies on the effect of brain lesions on growth because of interference with mechanisms controlling food intake, water balance, temperature regulation, and activity, as well as the secretion of several pituitary hormones. In an heroic effort to overcome these difficulties, Reichlin (1960) placed lesions in the hypothalamus of rats and gave vasopressin, testosterone, and thyroxine as replacement therapy, as well as employing pair-fed controls as a check against changes in the food intake of the lesioned animals, and found that growth was reduced by lesions in the anterior half of the median eminence which involved the arcuate nuclei and supraoptic-hypophysial tract. Others have focussed attention on the ventromedial nuclei in the rat, or have reported that lesions of the paraventricular nuclei in kittens depresses the rate of growth and that this effect could be reversed by treatment with growth hormone. In monkeys, the stimulatory effect of hypoglycaemia on GH secretion has been abolished by hypothalamic lesions, although the baseline levels of the hormone remained unchanged. Lesions in the hypothalamus may also interfere with the release of GH induced by capture or anaesthetization with ether in squirrel monkeys. Attempts to localize the hypothalamic receptors concerned with the control of GH secretion have also employed the microinfusion of glucose into the brain in an effort to block the release of somatotrophin in response to insulin. It was found that the response was reduced or abolished by an amount of glucose that had no effect upon peripheral glucose levels. Intrapituitary infusions of glucose did not block GH release but, unfortunately, the finding that ^{14}C-labelled glucose diffused from the median eminence to other parts of the brain precluded a precise localization of the receptors (Schalch and Reichlin, 1968).

The effect of electrical stimulation of the hypothalamus upon growth hormone secretion has not been greatly studied but the plasma concentration of this hormone in the monkey has been raised upon stimulation of the anterior median eminence, and has recently been shown to rise in the rat about 5 minutes after unilateral electrical stimulation of the ventromedial nucleus (Frohman, Bernadis, and Kant, 1968).

Growth Hormone Releasing Factor

In view of the novelty of studies of the neural control of growth hormone, the body of knowledge concerning growth hormone releasing factor is surprisingly large (see McCann and Porter, 1969). The first indication of the existence of a releasing factor for somatotrophin came from Deuben and Meites (1964), who showed that crude extracts of rat hypothalamus increased GH release from cultures of rat pituitary glands. Very quickly, this result was confirmed with extracts of pig or cattle hypothalamus (Schally, Steelman, and Bowers, 1965). Next, it was shown that the intracarotid injection of hypothalamic extracts in rats depleted the pituitary gland of GH (Pecile, Müller, Falconi, and Martini, 1965), and since then extracts from a range of mammalian species, as well as birds and amphibia, have been tested and found to be active in this way. Extracts of cerebral cortex of the appropriate species have been frequently used for control purposes and do not cause GH release.

Growth hormone releasing factor (GRF) activity (as assayed by measurement of the pituitary content of growth hormone in the rat after intracarotid injection) has been found to be lacking in hypothalamic extracts of rats injected with insulin, while pretreatment with reserpine prevented such a depletion. Stresses, such as exposure to cold, decreased hypothalamic GRF activity in rats, and this change was paralleled by a depletion in pituitary GH content and the appearance of GRF in the plasma. The plasma from rats hypophysectomized for 3 months has been reported to contain GRF activity which disappeared from the blood 10 minutes after occlusion of the cranial circulation, so implying that the activity came from the brain. Others have found that hypoglycaemia led to the appearance of GH-releasing activity in the plasma of hypophysectomized rats, and that the activity vanished after destruction of the median eminence. In kittens, bilateral lesions of the paraventricular nuclei, which caused a retardation of growth, resulted in the disappearance of GRF from the median eminence region. Retardation of growth in rats, brought about by the neonatal administration of cortisone, was associated with a lack of GRF in the hypothalamus, while the factor was present in specimens taken from controls. Such observations (Schally, Sawano, Müller, Arimura, Bowers, Redding, and Steelman, 1968) have been regarded as strong evidence in favour of the existence of a GRF, but a note of caution has crept into the discussion. In part, this is due to the discrepancy between the positive results based upon bioassay of the changes in growth hormone content of the rat pituitary gland, and the negative results which have stemmed from immuno-assay. In addition, preparations of stalk median eminence extract which have been found by several investigators to produce a marked depletion of pituitary GH when measured by bioassay, failed to elevate immunoassayable plasma GH levels or to consistently diminish the pituitary GH content in the rat (Schalch and Reichlin, 1968). It has been argued that the decreases in pituitary growth hormone found by bioassay must lead to enormous increases in the plasma GH level, which have not been observed. 'It is difficult to believe that such increases would escape detection by the radioimmunoassay, which, for example, has detected the elevated plasma GH levels of the fasted rabbit and mouse, and of the foetal rabbit' (Garcia and Geschwind, 1968). On this basis, the failure to demonstrate GRF effects either at the pituitary or plasma level of the rabbit and rat gains significance. However, Buse, Fulmer, Kansal, and Worthington (1970) have shown that blood plasma from the hypophysial portal vessels of normal rats raised the content of growth hormone in the pituitary gland of recipient rats after intracarotid or intravenous infusion. In the monkey, for which the immunoassay is well validated, GRF preparations which are highly active in the rat, are relatively ineffective. Crude GRF prepara-tions readily elevated plasma GH levels in monkeys but were shown to be contaminated with vasopressin. The presence of vasopressin does not satisfactorily account for all positive results and the situation has been further complicated by the results of studies in sheep, where an injection of an acidic extract of cattle stalk median eminence, free of vasopressin, caused a rise in the plasma level of GH (Machlin, Takahashi, Horino, Hertelendy, Gordon, and Kipnis, 1968). The bioassay for growth hormone releasing activity based upon the pituitary depletion of growth hormone in rats has encountered criticism because of inconsistent results, while preparations which apparently released growth hormone, as assessed by bioassay, were inactive in the immunoassay (Rodger, Beck, Burgus, and Guillemin, 1969).

Studies of the biological activities of growth hormone releasing factor, which have largely employed the bioassay for growth hormone, have been quite comprehensive (Schally, Sawano, Müller, Arimura, Bowers, Redding, and Steelman, 1968). The activity of the porcine material is abolished by incubation with trypsin, chymotrypsin, or pepsin, but, since the original material was not homogeneous, it is not certain that porcine GRF is a poly-peptide, although present results indicate that it may be an acidic polypeptide with a molecular weight of 2500. The material is highly potent in that 0·5 μg can deplete the growth hormone content of rat pituitary glands and is accompanied by a rise in plasma GH concentration, measured biologically. There is no effect on the secretion of thyroid stimulating hormone. A direct action of GRF on the pituitary gland is indicated by the observation that the depleting effect of GRF, or crude rat stalk median eminence extract, was not blocked by pretreatment of the assay animal with reserpine, chlorpromazine, Nembutal and morphine, or dexamethasone. On the other hand, chlorpromazine, reserpine, or dexamethasone did inhibit the release of GH caused by insulin, presumably by depressing the hypothalamus or a higher centre. In the light of the concept of the direct feedback action of pituitary hormones on the hypothalamus, it may be noted that the GH-releasing activity of highly purified or crude hypothalamic extracts can be blocked by pretreatment of the test rats with large doses (1 mg/100 g B.W.) of growth hormone. Transplantation of pituitary tumours secreting GH into rats has been associated with a decrease in the size of the anterior lobe of the hypophysis, in the number of acidophils, and in the concentration of GH. Implants of growth hormone in the median eminence and adjacent hypothalamic areas have also caused a fall in pituitary weight and in the concentration of endogenous hormone.

Since doses of releasing factor can discharge many times their own weight of hormone from pituitary tissue under *in vitro* conditions (the ratios reported range between 70 and a thousand million as purer preparations become available), the factor seems to promote the release and synthesis of somatotrophin. This view is supported by the fact that the action of GRF is blocked by actinomycin D, which interferes with the synthesis of proteins, while purified GRF does not alter the secretion of any other pituitary hormones in *in vitro* experiments. Schally and his co-workers have been so impressed with the actions of both porcine and bovine GRF in increasing the synthesis of GH, that they have renamed the principle the growth hormone releasing hormone (Schally, Arimura, Bowers, Kastin, Sawano, and Redding, 1968; p. 7). This step may be somewhat premature since evidence for the existence of a growth hormone inhibitory factor is coming to light, in that during the purification of growth hormone releasing factor from rat or sheep median eminence by gel filtration on a Sephadex column, certain fractions were found to reduce growth hormone secretion. A dose–response relationship could be established between the amount of substance added and the inhibition of growth hormone release from incubated pituitary tissue, and the fraction could also prevent the action of GRF, but not that of other releasing factors, on pituitary tissue. Because of the apparent specificity of action it was termed a growth hormone inhibitory factor (GIF) (Krulich, Dhariwal, and McCann, 1968; McCann and Porter, 1969).

10

Control of the Secretion of the Follicle-stimulating and Luteinizing Hormones

Because of the synergistic action of the follicle-stimulating and luteinizing hormones upon the ovaries, because of some uncertainty concerning the hormonal control of testicular function (p. 101), and because it is unlikely that either hormone is ever secreted alone, it is more satisfactory to discuss the control of secretion of these two gonadotrophins together than in separate chapters. In the case of the female, at least, division of the material could be achieved by separate discussion of the factors concerned with follicular development and with ovulation, as reflecting the activity mainly of follicle-stimulating hormone and then of luteinizing hormone, but then some repetition would be inescapable.

The Environment and Gonadotrophin Secretion

The sexual cycles of mammals are subject to modification by a wide variety of external influences acting through the nervous system. These can be grouped according to the length of exposure and the time taken to chart a response, with the latent period occupying months, days, and minutes. First come the external stimuli that determine the time of the onset of the breeding season in numerous species. Light is extremely important in this regard and the onset of reproductive activity in many species is coupled to changes in the length of the day. The ferret has been used in many laboratory experiments, for although the female normally comes into oestrus in the spring, sexual development can be advanced by exposing ferrets to artificially lengthened days from the autumn onwards. On the other hand, sexual activity in some breeds of sheep begins in the autumn and seems to be triggered by a reduction in the length of the day. No satisfactory explanation for this common response to diametrically opposite changes in day length is yet available (Donovan, 1966).

There is little doubt that light acts through the eyes to influence gonadotrophin secretion, for in the ferret, blinding prevents the ovarian response to prolonged illumination. It is certain that the neural stimulus from the eyes reaches the hypothalamus but the pathway travelled remains unknown. What seems clear is that beyond the optic chiasma the main optic pathway is not used, for damage to the optic tracts, or subsequent components of the visual system, fails to prevent the early onset of oestrus in ferrets exposed to prolonged illumination. Consideration of such a result has led to the suggestion that direct communications between the optic chiasma and hypothalamus exist. Such fibres have been described in histological sections of brain tissue by some anatomists but their occurrence has been as emphatically denied by others, so that this question remains unresolved. Despite the unlikely nature of the process, it is possible that light might have a direct action upon the hypothalamus, for it has been found that environmental light can penetrate the skull of the rat and sheep to reach the hypothalamus. One problem that has been little studied is the reason for the delay in the gonadal response of seasonally breeding animals to exposure to a suitably excitatory light-dark sequence. Usually about a month elapses before a sexually quiescent ferret comes into heat, yet only a fraction of the interval is needed for the gonadotrophin released to act on the ovaries to cause follicular development and oestrogen secretion. Nevertheless, a change in the electrical activity of the hypothalamus can be detected within a fraction of a second of exposure. Does the long delay indicate that the effect of light is cumulative? If it is, then how is the cumulative response stored within the brain?

Once sexual cycles have set in, as during the breeding season in sheep or during most of the year in rats and mice, environmental cues are utilized for other purposes. The alternation of day and night provides a timing mechanism for ovulation in the rat which has been studied in great detail. In the rat, ovulation recurs every four or five days in the absence of the male. The secretion of luteinizing hormone from the pituitary gland to provoke ovulation can be inhibited by injecting females with barbiturate drugs on the day before ovulation is expected, provided that the drug is given before 2 p.m. to animals exposed to 14 hours of light daily. Ovulation can again be delayed by repeating the treatment at the same time on each of the next two days but after this time regression of the follicles in the ovaries sets in and ovulation is no longer possible until a fresh crop ripens. Alteration of the lighting schedule can alter the critical period, though the effect is not immediate. When the light-dark schedule is revised, the circadian rhythm in the timing of ovulation in the rat undergoes a 12-hour shift but the resetting of the 'clock' takes about a fortnight to complete.

Other exteroceptive factors, besides that of light, are important. Sexual recognition and sexual behaviour in lower mammals is often promoted by means of olfactory stimuli stemming from the opposite sex (Bruce, 1970). The anosmic male goat is unable to select an oestrous female, while the odour of the boar is important in the immobilization reflex of the female which is essential for mating in the pig. Dogs are attracted by the scent of oestrous bitches. The onset of oestrus and ovulation is accelerated in sheep when a ram is added to a flock near the commencement of the breeding season and a similar effect has been observed in the goat. Olfactory stimuli are highly significant in the control of reproduction in the mouse. In the absence of males, female mice run irregular oestrous cycles and long periods of oestrus become common, but when a male is present the oestrous cycles become shorter and regular. The response to males is particularly striking when females previously grouped together are paired with stud males, for then a peak of mating activity is seen on the third night after the introduction of the male. This is the so-called Whitten effect, and can be elicited by permitting contact with the urine of male mice. While there is no doubt that the male provides a powerful stimulus it remains curious that removal of an individual from an all-female group to isolation favours the onset of oestrus in that female. Perhaps one female can suppress the secretion of gonadotrophin in another. Olfactory stimuli may also be much more important in primates than imagined hitherto. It has been found that the sexual performance of male rhesus monkeys varies with the phase of the menstrual cycle of the female, or in accord with the kind of sex hormone treatment given to ovariectomized females. The stimulus perceived by the male is probably olfactory because his sexual activity is increased by the application of minute amounts of oestrogen to the vagina which are in-effective when injected systemically, and is depressed by temporarily blocking the sense of smell (Michael and Keverne, 1968). Perfumes are used as sexual attractants in our own species and it may be significant that olfactory acuity varies in women during the sexual cycle and is greatest at the time of ovulation.

Auditory stimuli, provided by an electric alarm bell, have been shown to alter the oestrous cycle in rats and to induce phases of prolonged or persistent oestrus. Male rats, by contrast, remain unaffected.

Human sexual function is also profoundly influenced by the immediate environment, although much of the information available is anecdotal in character. Menstrual cycles often stop in young women placed in stressful situations and this condition is termed psychogenic amenorrhea. Shock, accident, or the loss of a close relative, as well as fear of pregnancy, are all known to disrupt the menstrual cycle, and the incidence of amenorrhea, menorrhagia, and irregular menses is known to be much higher among patients admitted to mental hospitals than among the general population. In other situations where there is an intense unsatisfied longing for a child, many of the signs associated with a pregnancy, such as breast development, amenorrhea and a steadily swelling abdomen may appear; these are cases of pseudocyesis.

The third group of neurally mediated influences affecting gonadotrophin secretion stem from sexual activity and provide classic examples of the neural control of the gonadotrophin secretion. Indeed, the fact that ovulation in the rabbit normally occurs as a consequence of coitus was recognized by Haighton in a paper to the Royal Society in 1797, although many years passed before the neuro-hormonal nature of this reflex was determined. In the rabbit, as in the cat, ferret, and other species in which ovulation does not occur spontaneously, reflex paths from the genital tract and from higher centres such as the cerebral cortex and amygdala to the hypothalamus are employed. Ovulation can thus be induced in the rabbit by artificial stimulation of the vagina and cervix but will also follow sexual excitement in females in which the genital tract and caudal part of the body has been rendered anaesthetic. After reaching the hypothalamus the stimulus evokes the discharge of the releasing factor for luteinizing hormone secretion and the gonadotrophin in turn causes ovulation.

Almost all of the examples illustrating the importance of the nervous system in the control of reproductive function mentioned above deal with the female. The apparent emphasis placed on female sexual function must not be taken to imply that the male is exempt from environmental influences, but simply that it is much more difficult to trace changes in testicular activity in the male. Testicular function in the males of seasonally breeding species waxes and wanes with changes in the length of day, and marked changes in testicular size can be plotted, but during the breeding season spermatogenesis and the secretion of androgen seems to remain constant. The oestrous cycle of the female rat is highly labile when compared with the apparently immutable production of sperm and the steady level of sexual activity shown by the male. However, the lack of change in the male may merely be a reflection of the coarseness of present indices of testicular action, for there are some subtle indications of a neural influence. Thus, it appears that coitus promotes the secretion of gonadotrophin in men, as shown by a subsequent increased excretion. There are also indications of the existence of an 8 to 10-day cycle in the excretion of oestrone and 17-oxosteroids in the human male, with the steroids coming from the testes (Exley and Corker, 1966). The basis of the rhythm is not known but may involve the feedback action of testicular hormones upon gonadotrophin secretion. Exley and Corker (1966) point out that peaks in the excretion of follicle-stimulating hormone and luteinizing hormone during the menstrual cycle with a 9-day interval have been described (though in work with biological, but not immunological, assay), as well as comparable maxima in the output of oestrogen, and suggest that the menstrual cycle is made up of three distinct cycles with a periodicity similar to that found in the male. The implied similarity in periodicity within the hormonal cycles of men and women may indicate a similar type of neural regulation of hormonal interplay in both sexes, but the existence of such rhythms needs first to be firmly established.

Many of the influences outlined above can be shown to affect the reproduction of individuals living in a community but, in addition, factors generated by the population itself can be very important. It is a remarkable fact that populations given an unlimited supply of food and apparently ample space do not increase in numbers to the extent which might be expected. As Snyder (1968) points out, in the house mouse, longevity, age at sexual maturity, the duration of breeding, frequency of pregnancy, intrauterine mortality, litter size, and neonatal mortality all vary according to population density. There is little doubt that an increase in ACTH secretion and in adrenal function can account in large part for the changes observed, particularly in cases where social dominance is important (chapter 7), but sexual function can be modified specifically. Testicular size and the sperm count in subordinate male mice is often low, while effects such as the pregnancy-block produced by a strange male could be operative in the female.

The Actions of Follicle-stimulating Hormone and Luteinizing Hormone

Although a variety of changes in sexual function are known to follow manipulation of the environment, the satisfactory interpretation of such changes in terms of alterations in the secretion of follicle-stimulating hormone and luteinizing hormone has proved difficult. This is because these two gonadotrophins show marked synergism in their actions and may never be secreted alone. When very highly purified, follicle-stimulating hormone (FSH) promotes development of the Graafian follicles in hypophysectomized females without the secretion of significant amounts of oestrogen. But when a small amount of luteinizing hormone (LH) is given together with a threshold dose of follicle-stimulating hormone then a greater enlargement of follicles occurs and oestrogen secretion is enhanced. In intact females, FSH causes more follicles to develop to maturity than normal so that an increased number of eggs is shed at ovulation. This is the effect which has complicated attempts to treat infertility by giving purified human follicle-stimulating hormone, so that some patients have given birth to four, five, or six babies. The action of FSH in the male is less spectacular, but the hormone has been shown to act on the first stages of spermatogenesis and to increase the size of the testes through enlargement of the seminiferous tubules. However, in immature males the development of mature spermatozoa is not hastened and there is no effect upon the interstitial cells (Lostroh, 1969).

Luteinizing hormone, when given to hypophysectomized female rats, reverses the atrophic changes evident in the steroid-producing interstitial tissue in the ovaries but does not elicit gonadal hormone release. But when some follicle-stimulating hormone is present, luteinizing hormone promotes the secretion of oestrogen and progesterone and enhances follicular growth and development. Ovulation occurs when luteinizing hormone is injected into females whose ovaries contain mature follicles, and corpora lutea are formed. The subsequent action of luteinizing hormone upon the corpora lutea is a matter of some dispute, for there are marked species differences; the secretion of progesterone from the corpora lutea is increased in the cow and rabbit, but not in the guinea-pig and rat. Luteinizing hormone may be more important than follicle-stimulating hormone in the support of testicular function in the adult male. In the hypophysectomized male mouse, for example, there is repair and stimulation of the testicular interstitial cells, with a parallel increase in weight of the accessory sexual organs, and spermatogenesis is resumed. The resumption of spermatogenesis may not directly result from an action of luteinizing hormone on the seminiferous tubules, for androgens alone maintain spermatogenesis in hypophysectomized males.

The very high degree of synergism in the actions of FSH and LH, and the fact that physiologically the two gonadotrophins never act alone, has complicated interpretation of the results of experiments in which the secretion of gonadotrophin is altered. After lesions in the hypothalamus in rats or guinea-pigs, ovulation may no longer occur regularly and it is clear that the secretion of luteinizing hormone has been disturbed. However, secretion of oestrogen in such animals continues, so the release of luteinizing hormone has by no means been arrested. Because of this difficulty the mixture of the two gonadotrophins that favours follicular growth and secretion has been termed 'folliculotrophin'. Although some have suggested that in reality follicle-stimulation, the secretion of gonadal hormones, and the induction of ovulation are different aspects of the action of a single hormone, it is undeniable that specific bioassays for follicle-stimulating and luteinizing hormone activity have revealed changes which move in opposite directions. Further, the isolation of one hypothalamic releasing factor for each gonadotrophin goes far to establish the separate identity of these hormones.

Gonadotrophin Secretion during the Menstrual and Oestrous Cycle

Information concerning the rate of secretion and changing blood level of a hormone is clearly essential before the factors controlling the release of that hormone are properly understood. To some extent this requirement can be evaded by tracing the changes in activity of the target organs influenced by the hormone but the use of such secondary indices is by no means satisfactory. Until very recently, biological assay methods were enployed for the measurement of the gonadotrophic hormone content of organs, body fluids, or urine, and these were so insensitive that extraction and concentration of the hormone was necessary before administration to the test animal. With the development of assay methods employing immunological techniques much has been gained in convenience, speed, and sensitivity, and it is now possible to measure the basal levels of a number of protein hormones in blood plasma. However, biological assays are by no means obsolescent. As yet, the use of immunoassays has been directed toward clinical problems because of the availability of highly purified human hormones for use as antigens, although comparable materials of animal origin are now becoming available. A haemagglutination-inhibition technique was first employed for the assay of human chorionic gonadotrophin, and while this procedure can be used for the measurement of luteinizing hormone, because of cross-reaction of the antiserum with the antiserum to HCG, it has been superseded by radioimmunological methods. Radioimmunoassays are about ten times more sensitive and are more accurate than haemagglutination reactions. An indication of the great gain in sensitivity derived from the application of immunological methods may be gleaned from work with follicle-stimulating hormone, where the standard is detectable at a level of less than 0·5 mµg, as compared with the 750–1000 mµg needed in biological assay. The plasma of normal men and women contains some 3 or 4 mµg per millilitre.

Current thinking concerning the secretion of gonadotrophin hormones by the pituitary gland is markedly different from that of a very few years ago. Then, the endocrine basis of the menstrual cycle was described along the following lines: At the time of menstruation, the blood levels of oestrogen and progesterone are low, alongside those of the gonadotrophins. During menstruation the secretion of follicle-stimulating hormone begins to rise, follicular growth in the ovaries is enhanced and the secretion of oestrogen also increases, probably by synergism between the actions of FSH and luteinizing hormone. This process continues until near ovulation when the blood level of follicle-stimulating hormone is high and a surge in the output of luteinizing hormone causes follicular rupture and corpus luteum formation. After this event the increased secretion of oestrogen and progesterone from the ovary inhibits gonadotrophin secretion and the release of these hormones falls away to reach a nadir at menstruation, when the cycle begins again. However, recent measurements of the blood levels of gonadotrophic hormones have failed to establish the occurrence of cyclic fluctuations in gonadotrophin secretion. Instead, the plasma concentration of follicle-stimulating hormone and luteinizing hormone remains practically constant, except for a short-lived peak of secretion of both hormones at ovulation, when the blood level of follicle-stimulating hormone rises some two and a half times and that of luteinizing hormone is sharply elevated to five times the basal secretion. Such peaks in hormone secretion would not be expected to occur on the basis of a purely hormonal interaction between the gonads and hypophysis and imply that the nervous system provides an impetus to hormone secretion at the time of ovulation. However, as Swerdloff and Odell (1968) point out, it is difficult to conceive of a slow (month long) central nervous system periodicity. If one existed, then occasional aberrantly timed peaks of LH might be expected and these were not seen in a series of more than fifty cycles studied. Further, if an inherent central nervous system rhythmicity exists, then study of post-menopausal or castrate women in whom the elevated FSH-LH concentrations in serum were suppressed to the same levels as observed in normal women should have revealed periodic peaks of FSH and LH. In studies lasting up to 60 days, no such peaks were observed. However, sequential administration of oestrogen and progestogen to postmenopausal or castrate women produced peaks of FSH and LH that closely resembled those detected in normal women and led to the conclusion that an ovarian signal determines the timing of the surge of FSH and LH for ovulation. The first sign of an impending ovulation seems to be a rise in the blood oestrogen level that then primes the neural mechanism leading to the burst in secretion of gonadotrophin (Harris and Naftolin, 1970; Donovan and Lockhart, 1970). But the basis of the increased secretion of oestrogen remains to be determined. In this connection, the diurnal rhythm in gonadotrophin secretion that is now evident in women may be of fundamental significance, although it is minor in degree when compared to the changes at ovulation.

Because of the difficulties involved in the measurement of minute amounts of gonadotrophins and gonadal steroids in the blood of laboratory animals there have been few studies of the changes in hormone secretion occurring during the oestrous cycle, although indirect indices such as the changes in the pituitary content of gonadotrophin or in uterine weight, or the ease with which gonadotrophin secretion can be blocked at different times of the day, have been used. The events taking place during the 4-day oestrous cycle of the rat have been summarized by Schwartz (1968a) as follows: On the day of oestrus a new set of follicles begins growing in the ovary, possibly under the influence of the FSH released along with the ovulatory surge of LH during the day before. Follicular growth continues during metoestrus, presumably due to the action of small amounts of FSH and LH. Luteinizing hormone is released between 11 a.m. and 3 p.m. on the day of dioestrus (day 3) to cause the output of oestrogen necessary during prooestrus, and again at about 2 p.m. on the day of prooestrus, along with FSH. As a result, the follicles begin the process of preovulatory swelling, oestrogen secretion stops and mating behaviour can be elicited. Some of the relationships between the levels of oestrogen, progesterone, and 20α-hydroxyprogesterone in ovarian venous blood and the systemic plasma levels of LH have been examined by Kobayashi, Hara, and Miyake (1968), who showed that the oestrogen content of the ovarian venous blood rose during the morning of prooestrus to reach a peak around 10 a.m. Luteinizing hormone became detectable in the plasma at 5 p.m. and its concentration rose rapidly to become maximal at 7 p.m., synchronously with the presence of progesterone in ovarian venous blood in greatest amount, after the release of oestrogen had declined. As with clinical studies, the surge in oestrogen release preceded that of LH. The Japanese workers suggested that the increased secretion of oestrogen from late dioestrus to the morning of prooestrus favours the biosynthesis or accumulation of LH by the hypophysis, as well as the secretion of FSH and a little LH, which in turn promotes the secretion of further oestrogen. More recently, antibodies to 17β-oestradiol and progesterone have been used to block the action of endogenous steroid. Ferin, Tempone, Zimmering, and Vande Wiele (1969) found that the administration of oestrogen antibodies on the 2nd day of dioestrus prevented the surge of LH and delayed the ovulation expected two days later, whereas comparable injections of antibodies to progesterone were ineffective. Later in the cycle, the sudden outpouring of LH on the day of prooestrus was preceded by a rise in the secretion of progesterone and 20α-hydroxyprogesterone (Goldman, Kamberi, Siiteri, and Porter, 1969).

Neural Control of Gonadotrophin Secretion

With the realization during the 1930s of the importance of the nervous system in the control of gonadal function, attempts were made to isolate the hypophysis from neural influence by cutting the pituitary stalk, although experiments involving stalk section were performed in dogs more than fifty years ago. The results of such experiments were generally unsatisfactory because of the variable changes in gonadal function that were observed. In rats, for example, the oestrous cycle remained normal, was lengthened, or was abolished; sometimes gonadal atrophy occurred, often the gonads remained normal. Order set in when the occurrence, or absence, of gonadotrophin secretion was related to the regeneration of vascular connections across the site of stalk section, and it was found that the amount of regeneration could be roughly correlated with the degree of reproductive activity that was resumed. Prevention of regeneration of the portal vessels by the insertion of a barrier between the brain and pituitary gland caused permanent gonadal atrophy. The precautions necessary in studies employing pituitary stalk section have been outlined earlier (chapter 3), so that it suffices to say that the controversy which raged over the effects of this operation in the middle 1950s has died away and it is generally agreed that the hypophysial portal vessels provide the functional link between the hypothalamus and pituitary gland. Similar consequences to those observed after complete and permanent destruction of the hypophysial portal vessels ensue after transplantation of the hypophysis away from the sella turcica, and are to be expected. The significance of these connections has been underlined by experiments in which the pituitary gland was transferred to a kidney and allowed to remain there for some weeks. Gonadal atrophy occurred, alongside hypofunction of several other target organs, but when the gland was replaced beneath the median eminence normal function was resumed.

The isolated pituitary gland secretes little follicle-stimulating or luteinizing hormone, and follicular activity in the ovaries of females stops. Under special circumstances, such as the implantation of several pituitary glands into one animal, the minimal secretion of each gland summates to reach a measurable level but the gonadotrophin released by a single gland exerts no detectable effect. Pituitary glands grafted to the kidney can be made to secrete gonadotrophin by the application of extracts of median eminence to them. On the other hand, the isolated hypophysis does secrete the third gonadotrophin, prolactin and, indeed, the release of this hormone seems to be favoured by isolation or transplantation of the gland. The control of prolactin secretion will be discussed later, but in the present context the continued release of this factor by transplanted pituitary tissue may account for the puzzling maintenance of testicular function in some males in which the pituitary gland has been moved away from the sella turcica, although it should be added that the role of prolactin in the control of testicular activity is by no means clear. It is also possible that the testis is more sensitive or responsive to gonadotrophin than is the ovary.

The secretion of the gonadotrophins is altered by electrical stimulation of the hypothalamus. The response is most striking in oestrous rabbits and cats, for ovulation results. After the pioneer investigations in which diffuse electric shocks were applied to the head, more precise techniques have been adopted and ovulation has been seen to occur in rabbits after stimulation of some part of the basal diencephalon-medial preoptic area, or of the tuberal or mamillary parts of the hypothalamus. A basal, tuberal area is particularly sensitive to stimulation but a response can be obtained from much of the hypothalamus. Much of the work on the rabbit has been confirmed in cats and it has been further found, in both species, that stimulation of the amygdaloid area in the temporal lobe causes ovulation. Excitation of the hypothalamus has also been found to hasten follicular rupture in the rat. Irritative lesions are useful in work of this kind and ovulation has been advanced by 24 hours in rats with five-day cycles by stimulation on the third day of dioestrus. Pseudopregnant rats have also been made to ovulate in this way. In other experiments on rats, ovulation has first been prevented by blockade of luteinizing hormone secretion with barbiturate drugs, and then triggered by electrical stimulation of the hypothalamus. Since ovulation can be induced by stimulation of the medial preoptic area, the anterior hypothalamic area, and the septal complex, it has been suggested that a diffuse system of neurones originates throughout the septal complex, converges as it enters the medial preoptic area and anterior hypothalamus, and assumes a restricted basal location as it reaches the tuber cinereum (Everett, 1964).

There are increasing indications of the existence of mechanisms inhibiting the secretion of LH. In one series of experiments, ovulation occurred in rats after stimulation of the amygdala on the second day of dioestrus. This response was blocked by section of the stria terminalis, which provides the major afferent pathway from the amygdala to the hypothalamus, although ovulation could still be induced after this operation by stimulation of the preoptic area (Velasco and Taleisnik, 1969). Stimulation of the amygdala has also been shown to cause a rise in the plasma level of LH, and it is of particular interest that concurrent stimulation of the hippocampus stopped the increased output of LH. Further, interruption of the fornices, which run from the hippocampus to the hypothalamus, enhanced the effect of amygdaloid stimulation. Stimulation of the ventral hippocampus on the day of prooestrus blocked spontaneous ovulation (Velasco and Taleisnik, 1969b). These results are somewhat similar to those obtained in work on the control of adrenal function, where stimulation of the hippocampus depressed the release of ACTH. However, Kawakami, Seto, Terasawa, and Yoshida (1967) and Kawakami, Seto, and Yoshida (1968) have different ideas. Studies of the EEG of oestrous rabbits, of rabbits after copulation, and of animals treated with progesterone or with LH, indicated that during oestrus the excitability of the hippocampus was lowered, while that of the amygdala was raised. Conversely, after coitus, treatment with progesterone, or with LH, the amplitude of slow (8–9 c/s) waves in the hippocampus was increased and its excitability considered to be enhanced, whereas that of the amygdala was depressed. Thus a reciprocal relationship appeared to exist between the activities of the hippocampus and amygdala. Nevertheless, ovulation occurred after electrical stimulation of either the amygdala or hippocampus. Since lesions in the dorsal fornix blocked the ovulation-inducing action of stimulation of the hippocampus, but not that of medial amygdaloid excitation, the response to hippocampal stimulation seemed to be specific and not due to spread to the amygdala. As illustrated in Figure 10.1, the Japanese investigators suggest that progesterone exerts a positive feedback action upon the hippocampus, and a negative feedback action upon the amygdala, although both neural structures facilitate the release of LH.

Most studies on the effect of stimulation of neural structures upon gonadotrophin secretion have employed the occurrence of ovulation as evidence of a positive effect. This is because ovulation provides an accurate and rapid response to a surge of luteinizing hormone and it is perhaps not suprising that attempts to influence the secretion of follicle-stimulating hormone by brain stimulation have been few. The induction of FSH-release in an individual with normally quiescent or inactive gonads may be a much slower process which may require chronic stimulation of the appropriate neural centres. Certainly little success in advancing the onset of oestrus in seasonally breeding females has been achieved by this means. It is also possible that the hypothalamic mechanisms controlling FSH secretion are subject to inhibition by other centres so that stimulation of the hypothalamus could be reinforcing the inhibition (Donovan and van der Werff ten Bosch, 1959a).

Recent work on the function of islands of hypothalamic tissue in experimental animals has shown that ovarian follicular activity is well maintained, so that incoming neural stimuli to the medial hypothalamus do not seem to be essential for the release of follicle-stimulating hormone. Surgical isolation of the medial hypothalamus can be achieved in a drastic fashion by removing the rest of the brain but it is now possible to obtain the same result with the aid of a special knife, the Halász knife, which is held in a stereotaxic machine and manipulated to make a cut around and above this part of the brain. Since the rest of the brain remains intact and undamaged, the animals survive extremely well and the results are much more meaningful. Less satisfactory results are obtained if the structures in the lateral hypothalamus concerned with eating and drinking are damaged (chapter 15).

In male rats the weight and histological structure of the testes are well maintained after deafferentation of the hypothalamus, but in females the oestrous cycles commonly stop after the operation, with constant follicular activity and vaginal oestrus resulting in some cases, and the occurrence of persistent corpora lutea and vaginal anoestrus being observed in others. To some degree, the differences in ovarian function after deafferentation of the hypothalamus can be correlated with the location of the anterior border of the cut, for present information indicates that if the suprachiasmatic nuclei are included then persistent vaginal cornification develops (Butler and Donovan, 1969). There is now little doubt that the anterior connections of the hypothalamus

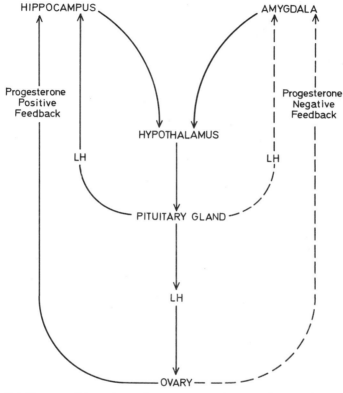

FIGURE 10.1 The possible actions of progesterone in controlling ovulation. In this view the hippocampus and amygdala operate antagonistically in their influence over the hypothalamus, and progesterone might promote the drive from the hippocampus and inhibit that of the amygdala. After Kawakami, Seto, Terasawa, and Yoshida (1967).

with the rest of the brain (largely through the preoptic area) are of greatest importance in the control of ovarian function, for division of these links, by making an anterior arc-shaped cut which extends a little caudally on each side of the hypothalamus brings about the same changes in ovarian function. Division of the posterior connections of the hypothalamus fails to disturb the oestrous cycle. These findings reinforce the conclusion drawn from studies of the effects of stimulation of the brain that afferent pathways reaching the hypothalamus through its anterior border are of prime importance in the control of ovulation, for if they are cut ovulation fails to occur, and that the hypothalamus possesses some autonomy in the control of gonadotrophin secretion.

When the functional capacity of hypothalamic islands is evaluated and related to the response to afferent stimuli, a two-level concept of the control of gonadotrophin secretion readily takes form. The basal level of secretion is mediated by the median eminence region of the hypothalamus and the constant, tonic, secretion of FSH and LH can be altered only under extreme circumstances such as ovariectomy, when castration cells develop in the pars distalis and

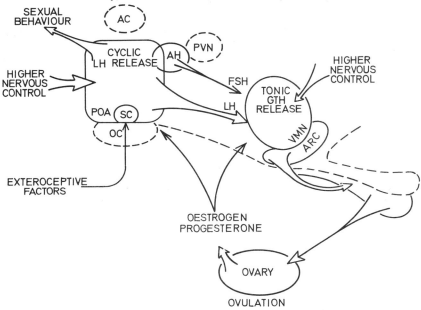

FIGURE 10.2 The control of gonadotrophic hormone secretion through the hypo-thalamus of the female rat, to show the two fundamental components: a tonic mechanism in the median eminence area, and a cyclic mechanism in the preoptic area. Abbreviations: AC, anterior commissure; AH, anterior hypothalamus; ARC, arcuate nucleus; OC, optic chiasma; POA, preoptic area; PVN, paraventricular nucleus; SC, suprachiasmatic nucleus; VMN, ventromedial nucleus. From Gorski (1966).

there is a rise in the content of LH. A trigger mechanism is superimposed upon the tonic mechanism to provide for the production of the surges in gonadotrophin secretion necessary for ovulation, and this is located in the preoptic area. An illustration of this hypothesis has been provided by Gorski (1966) and is presented in Figure 10.2. It will be seen that a variety of afferent neural influences are considered to modify gonadotrophin secretion through their action on the preoptic area, which in turn projects to the median eminence. Interruption of the projection would be expected to have similar consequences to those seen after the production of hypothalamic islands, and indeed such effects have been seen to follow the placement of electrolytic lesions in this area.

The ovarian changes seen in animals suffering hypothalamic lesions fall into three classes or syndromes. Lesions in the median eminence often produce atrophy of the gonads and accessory genitalia, probably because of interference with the discharge of releasing factors, whereas lesions in the anterior hypothalamus commonly abolish ovulation and lead to constant follicular activity in the ovaries. More dorsal lesions in the hypothalamus, at the border of the hypothalamus and thalamus above the paraventricular and dorsomedial nuclei, are associated with periods of prolonged vaginal dioestrus and the presence of more and larger corpora lutea in the ovaries than normal. It appears that FSH-RF may be synthesized in an area close to the paraventricular nuclei, while two areas, one in the suprachiasmatic-preoptic region and another in the arcuate-ventromedial region, are concerned with the production of LH-RF (Mess and Martini, 1968).

The Gonadal Hormones and Gonadotrophin Release

The effects of the gonadal steroid hormones upon gonadotrophin secretion have been actively studied for many years and provide the foundation of an extensive literature. In essence, the gonadal hormones, androgens, oestrogens, and progestogens, inhibit the secretion of the pituitary hormones when present in high concentration so that when the blood level of steroids falls, as after gonadectomy, the output of gonadotrophin rises. Problems arise, and curiosity is roused, when the interactions of gonadal and pituitary hormones are considered in detail, for on occasion, oestrogen or progesterone will evoke, not inhibit, gonadotrophin release. This is termed a positive feedback action. A single, suitably-timed injection of oestrogen causes ovulation in rats, but is ineffective after hypophysectomy, so indicating that the steroid provokes gonadotrophin secretion. Since this effect of oestrogen is blocked by drugs depressing neural activity, such as atropine, dibenamine, chlorpromazine, and the barbituates, or by hypothalamic lesions, and oestrogen implanted into the anterior hypothalamus enhances LH release, and the onset of puberty, there is little doubt that the gonadal hormone acts on the brain. Progesterone likewise can promote LH secretion (Mess and Martini, 1968).

The gonadal hormones do not exert parallel actions upon pituitary function, for oestrogen is much more powerful an inhibitor of gonadotrophin secretion than progesterone. Progesterone inhibits LH secretion to a much greater extent than that of FSH, so that follicular development in the ovaries can continue despite prolonged treatment with this steroid. Additional permutations are possible, for the steroids can act on the brain, or, to a much lesser extent, upon the hypophysis, and neuronal activity can be enhanced or suppressed, while the sensitivity of the nervous system as a whole to these hormones seems to vary cyclically. Attempts have been made to present the interactions between the gonads and the hypothalamo-hypophysial system in terms of control system theory and feedback servomechanisms, just as described for ACTH secretion, and two such examples appear in Figures 10.3 and 10.4, which are taken from the work of Schwartz (1968, 1969). The prime difference between the control of gonadotrophin and of ACTH secretion lies in the provision of a mechanism for the production of cycles of gonadotrophin secretion in the female. It is now quite clear that in the rat there is a tendency toward a daily surge in the release of LH between 2 p.m. and 4 p.m., which is great enough every 4th or 5th day to cause ovulation. Despite the existence of

109

diurnal variations, in the absence of an additional cyclic mechanism (or relaxation oscillation in control theory terms) a steady state with near-constant hormone levels would be reached. For relaxation oscillation, a process is required which, once triggered, runs to completion and requires recharging before triggering is again possible. Ovulation, with the conversion of an oestrogen-secreting follicle into a progesterone-secreting corpus luteum with a limited life-span, provides the biological equivalent of relaxation oscillation and the surge of LH provides the trigger. When ovulation is blocked (whether by brain lesions, or by constant illumination) a state of persistent oestrus is often observed and may be maintained for long periods. However, some question about the significance of ovulation remains, for sexual cycles occur in adolescent girls, and in sheep at the beginning of the breeding season, which are unaccompanied by ovulation. In formulating a model it is necessary to extract from a mass of physiological data those variables and interactions between components that are essential for the operation of the model and to express them meaningfully. Then it becomes possible to simulate the system on a computer and to mimic physiological experiments to see whether the results obtained with the model approximate to those determined with the living system. If the model is a close representation of reality, then 'experiments' can be carried out on the computer for comparison with future studies *in vivo* to test the accuracy of the predictions. Models tend to increase in complexity with time, as additional factors are introduced and this trend is illustrated by Figures 10.3 and 10.4. A rather diagrammatic example is provided as Figure 10.3 (Schwartz, 1968), and the simplest point at which to start an examination is with the effect of folliculotrophin (FSH+LH) upon oestrogen secretion and follicular growth (steps 1 and 2); an effect which is terminated by the surge in LH secretion that causes ovulation (steps 3 and 4) and the transformation of granulosa to luteal cells (step 5). The subsequent action of LH or prolactin upon the corpus luteum (step 6) to promote progesterone secretion (step 7) is controversial, as is the nature of the luteolytic mechanism (step 8) that destroys the corpus luteum in the absence of conception and varies from species to species. Both oestrogen from the follicles and progesterone from the corpora lutea exert a feedback action which is integrated with stimuli arising from the external environment (step 9). The rate of change of concentration of steroid hormones must be taken into account, since an abrupt rise in the oestrogen or progesterone level of the blood can cause ovulation, and is represented by d/dt, and these factors are shown to interact at the top left-hand corner of the diagram, where K_F stands for the set level of secretion of folliculotrophin (step 10) which can be depressed by oestrogen (step 11). The hypothalamic releasing factors intervene between the neural control system and the pituitary gland (step 12). The ovulatory surge of LH is the end result of changing concentrations of oestrogen (step 13) and progesterone (step 15) upon the setting for the rate of secretion of LH (K_L). Exteroceptive factors, such as the diurnal variation in light or the sensations associated with coitus, are also important (step 14) and act to enhance a positive drive to gonadotrophin secretion (step 17), which is limited by oestrogen and progesterone (steps 16 and 18). Additional complications enter into the control of prolactin secretion during the oestrous or menstrual cycle, but need not be discussed here.

The newest and most complex version of the model (Figure 10.4, Schwartz, 1969) takes the uterus and the neural centres concerned with mating behaviour

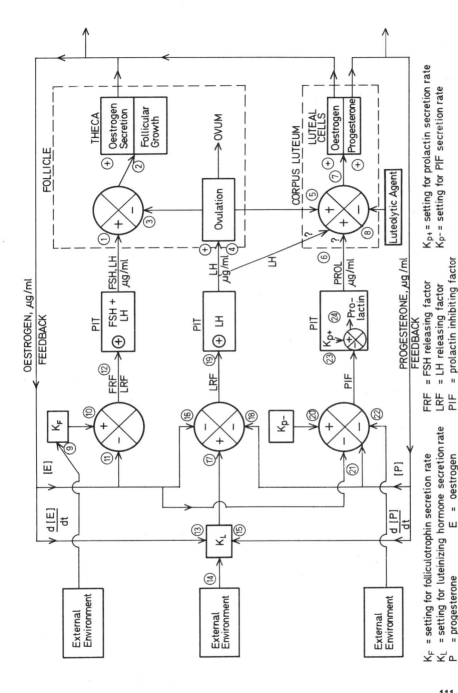

FIGURE 10.3 Regulation of ovarian hormone secretion and ovulation by a negative feedback–positive feedback relaxation–oscillation system in the rat. See text. From Schwartz (1968).

K_F = setting for folliculotrophin secretion rate
K_L = setting for luteinizing hormone secretion rate
P = progesterone

FRF = FSH releasing factor
LRF = LH releasing factor
PIF = prolactin inhibiting factor

E = oestrogen

K_{p+} = setting for prolactin secretion rate
K_{p-} = setting for PIF secretion rate

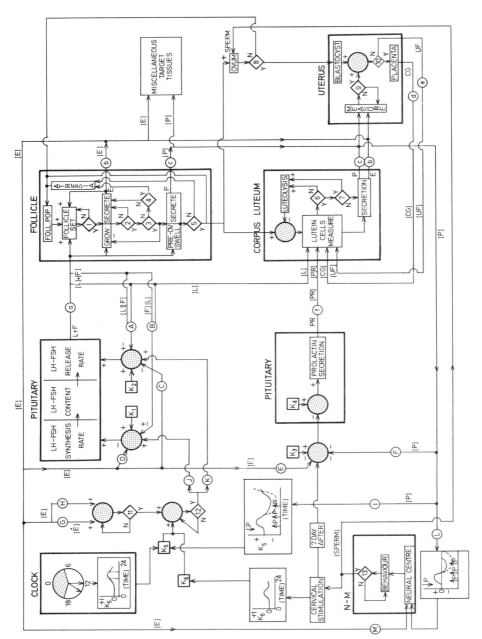

FIGURE 10.4 The Schwartz model for the control of the oestrous cycle of the rat. From Schwartz (1969).

112

THE SCHWARTZ MODEL FOR THE CONTROL OF THE RAT OESTROUS CYCLE

Explanation of symbols in model:

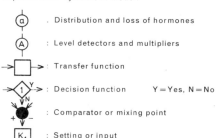

. Distribution and loss of hormones

: Level detectors and multipliers

: Transfer function

: Decision function $\quad$ Y = Yes, N = No

: Comparator or mixing point

: Setting or input

Explanation of Decision Functions:

1. Are LH and FSH both present?
2. Are follicles 'ready' to ovulate?
3. Is (LH) rising rapidly?
4. Have 3 or more days passed since answer 2 was yes?
5. Has ovulation occurred?
6. Are prolactin and/or chorionic gonadotrophin present?
7. Is LH or the 'uterine factor' (UF) present?
8. Has fertilization taken place?
9. Is the uterus prepared for implantation?
10. Did the blastocyst implant?
11. Are (E') and (E) in proper range?
12. Are K_5 and/or K_6 at threshold level?
13. Did mating occur?

Explanation of Settings (Inputs)

K_1 Setting(s) for LH and FSH rate(s) of synthesis.
K_2 Setting(s) for LH and FSH rate(s) of release.
K_3 Setting for factor inhibiting prolactin release.
K_4 Setting (pituitary) for prolactin release.
K_5 Clock input (modifiable by progesterone) for LH surge release.
K_6 Cervical stimulation input for LH surge release.

Explanation of Distribution and Loss Symbols

a Volume of distribution and time constants for LH and FSH.
b Volume of distribution and time constant for oestrogen.
c Volume of distribution and time constant for progesterone.
d Volume of distribution and time constant for chorionic gonadotrophin.
e Volume of distribution (direct channel uterus to ovary?) and time constant for uterine luteolytic factor.
f Volume of distribution and time constant for prolactin.

Explanation of Detectors and Multipliers

A   'Internal feedback' receptors for FSH and LH on release rates of FSH and LH.
B   As in A but for FSH and LH synthesis rates.
C   Negative feedback receptors and feedback constant for oestrogen levels on FSH and LH release.
D   As in C but for FSH and LH synthesis.
E   Feedback receptors for oestrogen on prolactin release.
F   Feedback receptors for progesterone on prolactin release.
G   Receptor and differentiator for oestrogen level for surge system.
H   Receptor for oestrogen level for surge system.
I   Receptor for progesterone for effect on clock signal (Ks).
J   Multiplier for surge system signal for FSH and LH synthesis rates.
K   Multiplier for surge system signal for FSH and LH release rates.
L   Receptor for progesterone for mating behaviour.
M   Receptor for oestrogen for mating behaviour.

113

into account and includes 'decision functions' (shown by the diamonds with Y for yes and N for no) to highlight the points of discontinuity or of switching. Other refinements have also been incorporated, but it is not appropriate to discuss the model in detail here, since it is given mainly to provide an example of current thinking. Full details will be found in Schwartz (1969), who also presents the case in favour of modelling. She considers that layouts like that of Figure 10.4 help to store and to organize information, and that they force the 'investigator to spell out hypotheses and concepts, rather than keeping these intuitive or implicit. This can be a shaking experience, but a valuable one'.

Most of the feedback action of gonadal hormones, whether positive or negative is exerted on the brain. The blockade of ovulation by drugs known to act upon the nervous system provided early evidence for this conclusion, which was supported by the finding that the advancement of ovulation by oestrogen or progesterone in rats could be blocked by anticholinergic or anti-adrenergic drugs. Some of the effects of lesions placed in the hypothalamus are believed to result from interference with the feedback action of oestrogen, for much higher doses of oestrogen than those normally effective were required to induce atrophy in rats and rabbits with hypothalamic lesions. Proof of the local action of gonadal hormone upon the brain has come from experiments involving the local implantation of steroid in selected areas. The rate of release of hormone from the depot must be low enough to avoid escape of detectable amounts of hormone into the systemic circulation and to avoid treatment of large areas of the brain, yet high enough to raise the concentration of hormone around the implant. At first, fragments of ovarian tissue were implanted into the hypothalamus of rats, with pieces of liver being used in control experiments, and it was found that only grafts of ovarian tissue located in the anterior hypothalamus inhibited gonadotrophin secretion, as shown by a significant fall in the weight of the uterus. In later work, oestrogen alone was used, and when applied to the arcuate-mamillary nuclear area of the hypothalamus caused long periods of dioestrus, loss of ovarian weight, and the appearance of the 'wheel' figures in the nuclei of ovarian interstitial cells that are usually linked with a deficiency of LH. Implants made into the suprachiasmatic area exerted little effect. The rise in plasma concentration of LH expected after ovariectomy in the rat was prevented by the implantation of small amounts of oestradiol into the basal tuberal hypothalamus but not by placement of steroid in the suprachiasmatic region or globus pallidus. Compensatory ovarian hypertrophy was similarly suppressed by implantation of oestradiol into the anterior hypothalamus or mamillary bodies, but not in the preoptic region, lateral hypothalamus, or amygdaloid area. Corresponding information is available for the rabbit but the results in rat and rabbit differ in that oestrogen applied to the mamillary bodies or anterior median eminence in the rabbit failed to alter the pituitary content of LH.

The feedback action of ovarian hormones in the female has been studied more intensively than that of the testicular steroids in the male. Nevertheless, testicular atrophy is known to follow the implantation of testosterone into the median eminence, while the placement of an anti-androgen, cyproterone acetate, was followed by enlargement of the accessory sex organs, so indicating interference with the feedback action of endogenous androgen (Davidson, 1967).

In order for experiments involving the direct application of oestrogen to the brain to have any physiological significance, it is necessary to prove that gonadal hormones can cross the blood-brain barrier to reach the hypothalamus. The fact that sexual behaviour, in which the brain plays an unquestioned part, is greatly influenced by the amount and class of gonadal steroids in the circulation can be taken to support this contention, but more direct proof has come from the use of radioactive oestrogen, autoradiography, and analysis of homogenized samples of tissue taken from ovariectomized cats and rats. The maximum concentration of radioactivity in control tissues such as muscle, thyroid, or posterior pituitary was attained within 2 hours, but the target organs for the oestrogen, such as anterior pituitary, uterine horns, and vagina, continued to accumulate radioactivity for 5–8 hours, and the radioactivity was retained for a further 6–12 hours. The hypothalamus and preoptic area, but not other parts of the brain, accumulated radioactivity in a similar way to the target organs and it seems that a bilaterally symmetrical neurological system involving the lateral septal area, preoptic region, and hypothalamus selectively takes up oestrogen from the blood. It is remarkable that the uptake of oestrogen by the brain was divorced from the manifestation of oestrous behaviour seen after treatment, for the peak of accumulation of radioactive oestrogen was reached after a few hours, but behavioural changes were not apparent for a few days, by which time all radioactivity within the brain had disappeared. Perhaps oestrogen serves to trigger metabolic changes which continue after loss of the steroid (Michael, 1965). In other work, the uptake of radioactive oestradiol by the hypothalamus, olfactory, and limbic structures of the rat brain was about one and a half times as great as the uptake of other regions. Two hours after the injection of a low dose of hormone (0·13 or 0·43 µg/100 g body weight) the content of radioactive material was still high in the limbic system and hypothalamus, but had fallen elsewhere (Pfaff, 1968). The particularly precise work of Stumpf (1968) with similar amounts of oestradiol has shown that the radioactivity became concentrated in the nuclei of neurones, but not in glial or ependymal cells, and that there was no difference in the location of activity in overiectomized mature female, immature female, or immature male rats. Labelled neurones were concentrated in the arcuate nuclei and the pars lateralis of the ventromedial nuclei. In the anterior hypothalamus the medial preoptic nuclei were heavily labelled and it appeared that the distribution pattern of the labelled neurones was largely identical with those areas wherein the various components of the stria terminalis end. That care is necessary in the evaluation of experiments involving the direct application of oestrogen to the brain or pituitary gland is illustrated by the work of Palka, Ramirez, and Sawyer (1966) who used tritium labelled oestradiol or oestradiol acetate. The steroid was fused into the bore of steel tubing and the needles introduced into the median eminence or anterior pituitary gland, with the location of the radioactivity later being determined by autoradiography and liquid scintillation counting of homogenized tissue samples. Detectable amounts of oestrogen could be found up to 2·0 mm away from the implant although the activity was very small. Tissue 0·5 mm away from the implant contained approximately 10% of the activity at the implant site, at 1·0 mm distant the figure had fallen to 1% and at 2·0 mm to 0·2%, or about 10^{-6} µg. After implantation of oestrogen into the median eminence greater amounts could be found in the ipsilateral half of the anterior pituitary gland than contralaterally, and there was some hypertrophy of the ipsilateral portion of the gland. The hypertrophy did not seem to be linked with the release of FSH or LH, but the output of prolactin may have been elevated. Implants made into the median eminence also have raised the plasma level of LH when this was measured five days postoperatively; thirteen days later LH was not detectable in the plasma.

Since oestradiol applied directly to the anterior pituitary gland also prevents a rise in the plasma concentration of LH after ovariectomy, and can modify the cytological changes seen in the hypophysis after spaying, it seems clear that oestrogen can act directly upon the cells of the pars distalis. This is also shown by the fact that hypertrophy of the ipsilateral lobe of the pars distalis occurred after the unilateral introduction of oestrogen into the gland of the rat. Mammary gland development, large corpora lutea, and persistent vaginal dioestrus, indicative of the secretion of prolactin, were demonstrable in these animals, but a rise in the plasma level of LH, seen in other rats with implants in the median eminence, did not occur (Palka, Ramirez, and Sawyer, 1966). Bogdanove (1963) has argued that many of the results of experiments involving local implantation of sex hormone can be explained on the basis of diffusion of the hormone into the hypophysial portal system whence it is distributed to much of the pars distalis. However, the inhibition of gonadotrophin secretion seen upon application of sex hormones to areas of the hypothalamus away from the median eminence, such as the mamillary bodies, is hardly explicable in this way, particularly as the implants may be very small and unilateral.

A practical application of information on the locus of the feedback action of gonadal hormones concerns the site at which oral contraceptives block ovulation. While the steroids would be expected to suppress gonadotrophin secretion through the hypothalamus, norethisterone in rabbits has been found to prevent the LH release expected after infusion into the pituitary gland of median eminence extract (Hilliard, Croxatto, Hayward, and Sawyer, 1966). The progestogen chlormadinone acetate blocked spontaneous or oestrogen-induced ovulation in rats after implantation in the basomedial hypothalamus, but for the blockade of ovulation in rats in which the median eminence was stimulated during dioestrus, it was necessary for the implants to be located in the anterior pituitary gland. Implants of chlormadinone located in the anterior hypothalamus did not prevent ovulation after electrical stimulation of the median eminence (Döcke, Dorner, and Voigt, 1968). It is always possible that oestrogen could backtrack from the pituitary gland to the hypothalamus but in the work on the rat only the mamillary region, which is of minor significance in this regard, could have been affected. Most recently, both progesterone and chlormadinone have been found to block the ovulation expected in the rabbit after the injection of median eminence extract into the pituitary gland (Spies, Stevens, Hilliard, and Sawyer, 1969). This result has been taken to indicate that the progestogens may act directly upon the pituitary gland, but could also arise from a central action which depresses the release of FSH (Harris and Sherratt, 1969).

Little is known about the changes in hypothalamic function brought about by the gonadal hormones. The size of the nucleus or nucleoli in the cells in some parts of the hypothalamus varies in response to gonadectomy or gonadal hormone administration but the changes have proved difficult to interpret. Thus, both spaying and oestrogen administration produce a reduction in the size of the nucleoli of the arcuate cell group although the volume of the nuclei of the cells is increased. An increase in the size of the nuclei in the cells of the preoptic area of male rats castrated on the day of birth has been observed

(Dorner and Staudt, 1968). Both oestrogen and progesterone lower the threshold to EEG arousal and the EEG after-reaction to coitus. Changes in the threshold to electrically-induced seizures upon treatment with gonadal steroids have also been described (chapter 16).

Direct Feedback Action of Gonadotrophic Hormones upon the Brain

Alongside the many observations of effects of gonadal hormones upon the brain, there are increasing indications that the gonadotrophic hormones may themselves act directly upon the nervous system, without the intermediation of the gonadal steroids. The first clues to the existence of this phenomenon came in the course of an investigation of the EEG after-reaction to coitus in the rabbit by Sawyer and Kawakami in 1959, in which a phase of sleep spindles lasting from several seconds to half an hour or more could be recorded, as well as a period of electrical hyperactivity in the hippocampus. The very rapid onset of this response implied that the after-reaction might be mediated by pituitary hormones acting back upon the nervous system, and substantiation of the suggestion came from experiments in which gonadotrophins induced a similar reaction in gonadectomized animals (so eliminating a possible feedback action of ovarian hormones) whereas ACTH gave negative results. Since then, FSH and LH have been implanted into the hypothalamus and seen to depress the pituitary content of endogenous gonadotrophin. There is a degree of specificity in the response because implants of FSH into the median eminence of rats depressed the content of FSH-releasing factor, when compared to sham-operated controls, but implants of LH into the median eminence or FSH applied to the amygdaloid areas was ineffective. The store of LH releasing factor in the median eminence normally rises in castrated and hypophysectomized rats, but the rise can be prevented by treating the animals postoperatively with LH.

Although it is difficult to challenge the many observations on the direct feedback action of gonadotrophins upon the brain (Motta, Fraschini, and Martini, 1969) the physiological significance of the effect remains open to question. If the mechanisms were operative then it becomes hard to explain why the production and release of gonadotrophin is so greatly increased after gonadectomy—for the elevated levels of pituitary hormone should suppress further release. However, those arguing in favour of the mechanism can point to the occurrence of changes in nucleolar size in some hypothalamic nuclei of hypophysectomized rats treated with ACTH after adrenalectomy, or FSH or LH after ovariectomy, whereas nucleolar size in these nuclei was unaffected by treatment of thyroidectomized-hypophysectomized rats with TSH (Ifft, 1966).

Control of the Onset of Puberty

Like most other aspects of gonadal function, the timing of puberty is under the control of the nervous system. This is shown by the many observations of the advancement of puberty by neurogenic stimuli applied to experimental animals. In the rat, for example, exposure to constant illumination or to handling in infancy hastens sexual maturation, while darkness or blinding has the opposite effect. In children in Europe, menarche (the first menstrual period) has appeared some 3 to 4 months earlier as each decade has passed, so that girls reach puberty about a year earlier than they did 30 years ago. A satisfactory explanation for this secular trend toward early puberty is not yet available. It is known that better-fed children tend to have an earlier menarche than poorly-fed girls but if nutritional factors provided the sole explanation then with rising standards of living the secular trend should have been arrested or slowed, and this has not been generally observed, although the trend toward early puberty in Scandinavia may have stopped. Since the number in a family has an effect on the timing of puberty, with menarche being delayed with increasing family size, subtler influences may be at work. Perhaps the general complex of sensory input is important so that the acceleration of sexual maturation in children may be due to the rich variety of stimuli provided by modern civilization. Experimentally, stressful stimuli have not been found to exert a consistent effect on the timing of puberty, which is sometimes advanced and sometimes delayed.

Clinical observations also point to the brain as being concerned with the timing of puberty, although they have not contributed greatly to present understanding of the processes concerned. Precocious puberty in children is very often associated with a variety of cerebral diseases, neoplasms and tumours, or has no apparent cause, when it is termed cryptogenic. What is important is that, apart from cryptogenic cases, only intracranial changes give rise to true precocious puberty, when ovarian or testicular activity fully matches that of the adult and pregnancy becomes possible in the infant girl or the boy produces mature spermatozoa.

An instance of a sexually precocious boy of 3 years is mentioned by Pliny, while a clinical case was described by Mandeslo in 1658. This was a girl whose menstrual periods began at 3 years and gave birth to a son at the age of six. Some one hundred years later the tale of Anna Mummenthaler was told by von Haller. Menstruation began in this girl at two years of age, she experienced a still birth at nine, the menopause at fifty-two, and died at the age of seventy-five. The cause of the sexual precocity of Anna Mummenthaler is not known, but it seems evident that any pathological changes in the brain must have been of the mildest kind. Study of the cases available at the beginning of this century indicated that the pineal gland was concerned in the control of sexual maturation, for pubertas praecox most commonly was seen when tumours of the pineal gland, or in that neighbourhood, were present. Many attempts were then made to study the function of the pineal gland in experimental animals in the light of the idea that the organ was responsible for the inhibition of sexual maturation in infancy, but few convincing results were obtained. It remains an intriguing but unexplained fact that sexual precocity in association with pineal tumours is very much more common in boys than in girls.

Tumours in the pineal region can give rise to pressure effects within the brain so that structures at a distance can be affected, and since brain tumours can cause precocious puberty without involving the pineal gland it is reasonable to presume that the function of the hypothalamus in infancy is altered. Some 30 years ago it was suggested that the posterior hypothalamus served to inhibit the gonadotrophin-stimulating activity of the anterior hypothalamus, and that damage to the posterior hypothalamus freed the anterior hypothalamus from inhibition with consequent accelerated release of gonadotrophic hormones from the anterior pituitary. While it is true that precocious puberty in children is seen most commonly in cases which involve the posterior hypothalamus, the physiological evidence in mammals other than man in favour of an inhibitory influence from the posterior hypothalamus on the control of gonadotrophin secretion is slender indeed. On the other hand, no satisfactory alternative explanation for the clinical cases of sexual precocity has yet been put forward.

Several aspects of gonadal function in infancy must be taken into consideration in any discussion of the mechanisms involved in the timing of puberty. The first point to be made is that gonadal function in infancy is by no means negligible. The secretion of gonadal steroids by the ovary is sufficient to keep the release of gonadotrophin in check, as can be demonstrated in a variety of ways. Removal of the ovaries in infant rats causes the appearance of 'castration' cells in the hypophysis but this change can be prevented by the administration of minute amounts of oestrogen—much less than that needed by the adult when comparison is made on a body weight basis, and about one-third of the dose necessary to cause growth of the uterus. After gonadectomy in infancy the gonadotrophin content of the pituitary gland rises, and the hormone becomes detectable in the blood, unless replacement therapy with oestrogen is instituted. Similar considerations apply to the immature male. Hemicastration of calves caused enlargement of the remaining testis which was accompanied by an increased secretion of androgen. This observation, besides indicating that the infant gonad secretes hormones, also shows that a similar array of gonad-pituitary interactions is demonstrable in infancy as in adult life. Why then does puberty set in, and what upsets the balance existing between gonadal hormone secretion and hypophysial hormone secretion in infancy in favour of an increased secretion of gonadotrophins at adolescence?

Current concepts relevant to the timing of puberty (Donovan and van der Werff ten Bosch, 1965; Critchlow and Bar-Sela, 1967) take very close account of the information available on the control of sexual function in the adult, for the mechanisms involved are similar. Indeed, the control of sexual development is just one activity governed by the hypothalamic mechanisms described earlier.

As long ago as 1932, Hohlweg and Junkmann considered that a sexual centre in the brain was responsible for the inhibition of gonadotrophin secretion during infancy and that the centre was exquisitely sensitive to oestrogen. With increasing age, and the approach of puberty, the centre became progressively less sensitive to oestrogen so that the release of increasing amounts of gonadotrophin was allowed. There is now no doubt that oestrogen acts on the hypothalamus, and perhaps elsewhere in the brain, during infancy. In infant female rats, short-term treatment with oestrogen near the expected time of puberty hastens this event. Further work has established that oestrogen acts on the preoptic area, for local application of oestradiol to this region for 48 hours in 26-day-old rats advanced vaginal opening and the early initiation of oestrous cycles. When oestradiol was left in the brain for longer periods the secretion of pituitary hormone was suppressed and ovarian development inhibited (Smith and Davidson, 1968). Since lesions placed in the anterior hypothalamus of infantile rats also advance the onset of puberty (Donovan and van der Werff ten Bosch, 1959b) it can be argued that the lesions interfere with the feedback action of oestrogen and free more caudal parts of the hypothalamus or median eminence from inhibition. While plausible, this explanation certainly begs many questions, for others have found that local implants of oestrogen in the ventromedial-arcuate nucleus region of the hypothalamus, or to the median eminence, also causes precocious vaginal opening. Smith and Davidson, for example, suggest that it is the threshold of a 'negative feedback controller' in the median eminence that increases during development and that the resultant gradual increase in the circulating levels of oestrogen activates a 'positive feedback controller' in the anterior hypothalamus, which in turn promotes the secretion of gonadotrophin in sufficient quantity to precipitate puberty. Experiments of quite a different kind have provided additional evidence of a positive feedback action of oestrogen or progesterone upon the brain during infancy. Not only will oestrogen cause ovulation in the adolescent rat (the Hohlweg effect) but injections of pregnant mare serum gonadotrophin given to 24-day-old rats elicit ovulation some two and a half days later. The ovulation can be blocked by neurally active drugs (barbiturates, atropine, or chlorpromazine) given at about 1 p.m. two days after the gonadotrophin injection, but if administration of the blocking drug is delayed until 4.30 p.m. there is no interference with ovulation. These findings indicate that there is a cyclic drive toward gonadotrophin secretion in infancy, just as in adult life, and that the cyclic mechanism is sensitive to the feedback action of gonadal hormones only at certain times of the day.

Explanations for the timing of puberty along the lines set out above are oversimplifications because the activities of extra-hypothalamic areas are neglected, and the possibility that hypothalamic activity and gonadotrophin secretion may be inhibited by another part of the brain is ignored. The latter possibility is a real one, for isolation of the medial basal hypothalamus in 3-week-old rats has been found to advance vaginal opening, although ovulation does not occur, and it is reasonable to suppose that inhibitory pathways to the hypothalamus have been severed. The inhibitory influence may arise in the hippocampus and be antagonistic to an excitatory effect arising from the amygdala—a situation which may have its parallel in the mechanism controlling ovulation in the adult (p. 106). Accordingly, lesions in the ventral hippocampus allow the induction of ovulation by PMS in the rat at 22 days. whereas the earliest response in animals without lesions occurs at 28 days. Despite the attractive nature of this concept, contradictory evidence cannot be ignored, for electrolytic lesions in the amygdala have been shown to advance puberty, as have lesions in the stria terminalis, whereas electrical stimulation of the amygdala or lesions in the hippocampus have delayed vaginal opening in the rat. Although it has been suggested that the lesions (made with steel electrodes) exerted an irritative effect and so were stimulatory in character, it is hard to account for the delay in puberty seen with electrical stimulation of the amygdaloid area on this basis. There is a further complication in that the implantation of oestradiol into the habenular region of 26-day-old female rats has been found to delay the onset of puberty and to act in opposite fashion to similar implants in the median eminence (Motta, Fraschini, Guiliani, and Martini, 1968). As an index of the sensitivity of the brain to gonadal hormones, the threshold to electrically-induced seizures in limbic structures has been followed during infancy and sexual maturation in rats. A decrease in the seizure threshold of the amygdala occurred just before puberty, which may have been due to oestrogen action, whereas there was no significant change in the threshold of the hippocampus (Terasawa and Timiras, 1968a).

Mere identification of the brain structures concerned in the control of gonadotrophin secretion during infancy would not greatly advance understanding of the processes involved, for immediately more fundamental questions arise. If the brake on gonadotrophin secretion in infancy resides in the hippocampus, how might the hippocampal brake be removed? What subtle changes in neuronal function are responsible for the shift in hippocampal activity? In what way do the gonadal steroids influence brain function? It is a measure of present ignorance that the word hippocampus may be replaced by hypothalamus or amygdala without loss of meaning.

Little mention has been made of the male in discussing the physiology of puberty. Although precocious puberty in boys is well known, it has proved difficult to advance puberty experimentally in male laboratory animals, even by direct administration of gonadotrophin. However, the local application of an anti-androgen to the hypothalamus of young male rats enhanced growth of the reproductive system although spermatogenesis was not hastened. Conversely, the introduction of testosterone into the hypothalamus depressed testicular weight in prepuberal animals, so that a negative feedback action of testicular androgen upon the hypothalamus is apparent during infancy in the male rat. At least for the rat, true precocious puberty, with the presence of fertile sperm, is not to be expected for the time period before puberty is very little greater than that needed for the completion of spermatogenesis—which requires about 40 days. Species with a longer period of infancy are better employed in studies of the control of puberty in the male.

The Releasing Factors

The occurrence of an hypothalamic factor for the release of follicle-stimulating hormone (FSH-RF) was first demonstrated by Igarashi and McCann (1964) some six years ago. It caused an increase in the plasma concentration of FSH in spayed rats in which FSH release had been previously blocked by treatment with gonadal steroids or by hypothalamic lesions. In the same year (1964) Mittler and Meites found that extracts of rat hypothalamus initiated FSH secretion when they were applied to cultured pituitary tissue, while extracts of cerebral cortex were not potent. Castration of male donor rats a fortnight before the collection of the median eminences raised the content of FSH-RF and, conversely, treatment of the donor animals with testosterone propionate for two weeks before death, depressed the FSH-RF activity of the extracts. Similar observations have been made upon the feedback action of ovarian hormones on the FSH-RF content of the median eminence of the female (Mess and Martini, 1968). Thus the concentration of this releasing factor in the median eminence can be altered by experimental procedures and the change in concentration is related to the level of secretion of the pituitary hormone. It is significant that FSH-RF becomes demonstrable in the plasma of male rats two months after hypophysectomy, and that long-term treatment with testosterone propionate reduced the plasma content of the releasing factor to undetectable levels (Negro-Vilar, Dickerman, and Meites, 1968). This observation provides evidence of the clearest kind of a feedback action of the steroid upon the brain. The pituitary content of FSH in intact rats has been depressed upon the administration of FSH-RF. The relationship between the amount of releasing factor added to a culture of pituitary tissue and the quantity of hormone released has been studied recently in some detail (Jutisz, 1967), and a log-dose response curve has been drawn between the amount of mediator added and the FSH released, so that 0·05 µg FSH-RF/mg pituitary tissue releases about 4 µg FSH/mg pituitary. Current information indicates that the molecular weight of FSH-RF may not be larger than 300 (Schally, Arimura, Bowers, Kastin, Sawano, and Redding, 1968). The substance appears to be a polyamine derivative, but since it is available only in minute quantity complete synthesis may be necessary before full studies of biological activity become possible. Nevertheless, the factor is active *in vivo* at doses of a few nanograms.

Work with FSH-RH has proved slow and laborious because of the lack of a simple, sensitive, and accurate assay for follicle-stimulating hormone. By contrast, discharge of LH is readily detected by the occurrence of ovulation in suitable circumstances, such as in the rabbit, where this event normally only follows coitus. Much early work was based on the ovulation test, and this continues to be of value, but the popularity of the ovarian ascorbic acid depletion assay for LH performed in the pseudopregnant young rat has also facilitated progress. Campbell, Feuer, and Harris (1964) found that the slow infusion of an acid extract of median eminence directly into the pituitary gland of a rabbit caused ovulation and that 20–50 times the dose effective when applied to the hypophysis was necessary when the intravenous route was used. Infusion of crude median eminence extract into the pituitary gland of the rat caused ovulation in animals in which the neural drive to gonado-trophin secretion was blocked by treatment with a barbiturate drug. The concentration of LH in the blood of rats can be raised by administration of LH-RF, whereas LH continues to be undetectable in the blood of treated hypophysectomized rats. The content of LH-RF has been shown to be elevated in portal vessel blood collected from hypophysectomized rats.

Some studies on the changing concentration of LH-RF in the hypothalamus during different reproductive states have been carried out. The content of stored releasing factor declines at prooestrus in the rat although the pituitary gland is similarly sensitive to the action of the releasing factor at all stages of the oestrous cycle. Further, the hypothalamus of the immature rat contains as much LH-RF as that of the adult animal, so that the secretion of gonado-trophic hormone seems to reflect changes in the store of releasing factor. Accordingly, large doses of testosterone reduce the content of LH-RF in the hypothalamus of the rat although the action of the factor on the pituitary gland is not impeded (McCann and Porter, 1969).

At one time, the secretion of LH was believed to be controlled by an adren-ergic agent but this view has since been discarded, although adrenergic, cholinergic, and even dopaminergic synapses may be employed in the path-ways from the periphery to the median eminence. Other substances found in the median eminence, including α-melanocyte-stimulating hormone, β-melanocyte-stimulating hormone, vasopressin, oxytocin, histamine, acetyl-choline, substance P, and bradykinin have been tested for the ability to provoke LH release but have been found to be inactive. Until very recently, LH-RF was believed to be a small polypeptide which was heat stable. In accord with this view its biological activity could be destroyed by trypsin or pepsin, or by hydrolysis with hydrochloric acid, but it now seems that the active fraction contains too small a percentage of amino acids to be a polypeptide. Both FSH-RF and LH-RF have been demonstrated in human hypothalamus as well as in that of other mammalian species examined, but difficulties have been experienced in completely separating FSH-RF from LH-RF in extracts of hypothalamic tissue and until this is done the separate identity of the two substances will remain uncertain. However, success has been reported recently with material of porcine origin.

As in the case of FSH-RF, a linear response in the release of LH as a function of the log-dose of LH-RF has been observed. In work with porcine LH-RF given to rats pretreated with oestrogen and progesterone, amounts as low as 0·002–0·005 μg were active and it was calculated that LH-RF released 5–10 times its own weight of LH. Other investigators have employed tissue culture methods and similarly concluded that 1·2 μg LH-RF added to the medium released 5 μg LH, which approximated to twice the initial content of the hormone in the pituitary tissue and indicates that the mediator does more than just cause discharge of hormone.

With the greater availability of releasing factors the analysis of their mode of action on the pituitary gland is underway. Work carried out with cultivated pituitary tissue treated with FSH-RF has shown that loss of FSH into the medium was not accompanied by a fall in the FSH content of the gland, so that there was a net production of gonadotrophin. The action of LH-RF on the pituitary gland can be blocked to a large extent by puromycin (which inhibits the biosynthesis of peptide chains). This result also implies that the releasing factor favours the synthesis of luteinizing hormone, but it is not yet certain whether these effects follow indirectly from a fall in the concentration of stored hormone in the gland. Nevertheless, it is again evident that the term 'releasing factor' focusses attention on but one aspect of the action of the mediator.

11

Control of the Secretion of Prolactin

Various names have been given to the pituitary hormone here called prolactin. These are recorded by Lyons and Dixon (1966), who prefer the term mammotrophic hormone in view of its prominent effect upon the mammary gland. Luteotrophic hormone is a term that has also enjoyed popularity because of the action of the hormone in promoting the secretion of progesterone by the corpus luteum of the rat. But since the corpora lutea of only a few species seem to respond in this way, and because luteinizing hormone can also be luteotrophic, the term is best avoided. The use of the word 'prolactin' does not imply that the activities of the hormone are restricted to the mammary gland but can extend to other functions allied to the promotion of lactation.

Prolactin is an important factor in the induction of lactation but does not act alone. Oestrogen and progesterone are needed for the development of the glandular apparatus of the breast and to ensure the responsiveness of the gland to prolactin. Growth hormone, adrenocorticotrophic hormone, insulin, and thyroxine act to enhance the effects of prolactin, while oxytocin is necessary to ensure the expulsion of milk from the gland to the young. However, a combination of prolactin and adrenocortical hormone is sufficient to induce lactation in the hypophysectomized guinea-pig or rat, and sheep prolactin alone maintained lactation in the hypophysectomized rabbit (Cowie, Hartmann, and Turvey, 1969). Prolactin acts directly upon the crop sac of the pigeon to cause growth of the organ and the formation of the 'crop milk' used for the feeding of the squabs. Local injection of prolactin into the crop sac causes a localized thickening of the wall and this response has provided the basis of the widely used pigeon crop sac assay for prolactin. Local injection of prolactin into a mammary gland of the female mammal also stimulates growth and lactation in an area or sector of the organ (Lyons and Dixon, 1966).

Prolactin has long been known to be concerned with the function of the corpora lutea in the rat and it has been thought that the hormone exerted a similar action in other species. Unfortunately, samples of prolactin that are highly luteotrophic in the rat, mouse, and ferret are not necessarily so in the rabbit, guinea-pig, hamster, cow, sheep, or pig. The human corpus luteum also does not respond to prolactin like that of the rat or mouse. Other mechanisms, not involving prolactin, that are important in supporting corpus luteum function include the action of LH—or a mixture of FSH and LH—as a luteotrophic hormone, and the operation of a corpus luteum-destroying, or luteolytic, substance produced by the uterus. Descriptions of these mechanisms have been provided by Rothchild, 1965; Bland and Donovan, 1966; and Donovan, 1967, among others.

There is as yet no certainty that the human pituitary gland secretes prolactin, for highly purified samples of human growth hormone possess prolactin-like activity (p. 90). Since the ratio of growth-promoting to lactogenic activity can vary from preparation to preparation, it is argued that two separable hormones exist. In this connection it may be noted that organ cultures of human foetal pituitary tissue produce steadily smaller amounts of immunologically measured growth hormone, while prolactin production rises. It would be a little surprising if the situation in the primate differed markedly from that of other mammals and it may be significant that growth hormone releasing activity is present in the human hypothalamus alongside prolactin inhibiting factor.

The stimulus of suckling is important in modifying pituitary function in the female (Donovan and Harris, 1966; Everett, 1966). Recent experiments have shown that the pregnant rat licks her genitalia and nipple lines more than under non-pregnant conditions and that the neural stimulation may promote mammary gland development. Prevention of self-licking by fitting the females with collars, retarded the growth of mammary glands. Suckling has a rapid action in lactating females in causing oxytocin release from the neuro-hypophysis for milk ejection (p. 60), as well as the secretion of prolactin and growth hormone. When applied to non-pregnant animals, as in experiments in which rats were given a series of litters to nurse, the oestrous cycles stop, dioestrus and a probable pseudopregnancy appears, and lactation sets in. The number of pups in a litter is important, for the intensity of the effect is related to the number of mammary glands suckled. In cows, too, the interval from parturition to first oestrus has been shortened by reduction in the frequency of milking. The restraint on the secretion of gonadotrophic hormone which is apparent during phases of prolactin secretion is thus eased. A similar phenomenon is known in women, for lactation may be prolonged for many months if weaning is delayed. Breast stimulation by suckling has caused mammary development and secretion in virgin girls, non-pregnant women, and, occasionally, men. Injections of prolactin, or of adrenal corticoids, retard mammary involution in non-suckled rats, and suckling has been shown to cause prolactin release, so that there is little doubt that prolactin secretion is favoured by suckling. The suckling or milking stimulus is not essential for the maintenance of lactation in the sheep or goat, as demonstrated by experiments involving transection of the spinal cord or, more strikingly, transplantation of a mammary gland to the neck. On the other hand, milking of virgin, spayed goats will cause mammogenesis and elicit lactation so that stimuli from the mammary gland remain quite significant and may be fully utilized in the intact female. In rabbits, lactation will continue in animals anaesthetized during the daily nursing period. Other exteroceptive factors can be made use of, as shown by the fact that the prolactin content of rat pituitary glands falls when the animals are allowed to see their litters.

Attempts have been made to trace the neural pathway from the nipples to the hypothalamus. Transection of the spinal cord in the rat, with suckling allowed at the anaesthetic nipples, inhibited lactation, but if suckling was allowed at the sensitive thoracic nipples, lactation continued. The tracts employed within the spinal cord are not clear and it seems that the route may be indirect, utilize small nerve fibres, and be widely dispersed through the reticular formation (Tindal, 1967).

In rats and mice, prolactin prolongs the life of the corpora lutea and induces pseudopregnancy. Sterile coitus, or mechanical or electrical stimulation of the uterine cervix also causes pseudopregnancy, and hence prolactin release. Prolactin secretion may go on for about a week (Zeilmaker, 1965). Stimulation of the cervix in the rat also causes a fall in the concentration of prolactin in the pituitary gland. The induction of pseudopregnancy by cervical stimulation is blocked by division of the afferent pelvic nerves or by spinal anaesthesia, but sterile copulation is still effective. Thus there may be several kinds of input to the brain that can evoke prolactin secretion (Everett, 1966). Olfactory stimuli provide one form, the visual stimulus provided by the young, another. Irritation of the nasal mucosa in the rat can induce pseudopregnancy, like other stressful stimuli, and ablation of the olfactory bulbs lowered the incidence of spontaneous pseudopregnancies in mice. The sense of smell may be concerned with the prevention of prolactin secretion, for the interruption of pregnancy caused by the presence of a strange male (the Bruce effect) can be overcome by the injection of prolactin or the insertion of a homograft of pituitary tissue. Seemingly, the odour of a strange male is effective in stopping or depressing the release of prolactin.

Many drugs have been given to rats to determine whether pseudopregnancy could be induced, or blocked, by pharmacological means, but no consistent picture has emerged except in the case of the tranquillizing agents, where it has been demonstrated that chlorpromazine and reserpine depress gonadotrophin secretion, with interruption of oestrous or menstrual cycles, while enhancing the release of prolactin to cause mammary development and lactation (Donovan and Harris, 1966). In view of the action of these drugs on the brain, the secretion of prolactin would be expected to be changed through an action on the nervous system, but a direct action on the pituitary gland has not been completely excluded. Reserpine inserted into the basal tuberal region of the hypothalamus of oestrogen-primed spayed rabbits induced lactogenesis, but was not active when directly applied to the pituitary gland. Systemically injected reserpine was similarly ineffective if the posterior tuberal region of the hypothalamus of the test rabbits was damaged.

The Inhibition of Prolactin Secretion

Despite the apparently stimulatory action of cervical stimulation or suckling in eliciting prolactin release, but in accord with the action of the tranquillizing drugs, which depress neural activity, it is clear that the secretion of prolactin is inhibited by the hypothalamus. Afferent stimuli causing prolactin release probably operate by removing the inhibition, and accordingly pituitary tissue disconnected from the hypothalamus, releases prolactin in increased quantity.

Complete interruption of the pituitary stalk in pseudopregnant rats and ferrets favours maintenance of the corpora lutea through the continued release of prolactin. Lactation in rabbits is also not nearly so depressed after pituitary stalk section as after hypophysectomy, and some of the disturbance that does follow can be attributed to a shortage of pituitary hormones, other than prolactin, which are important metabolically (Donovan and van der Werff ten Bosch, 1957; Cowie, 1969). In the stalk-sectioned goat, milk yield was depressed to about 30% of normal, but rose when growth hormone, tri-iodo-thyronine, insulin, and adrenal cortical hormone replacement therapy was given. Prolactin secretion also occurs in women after interruption of the pituitary stalk for the relief of metastatic mammary cancer, as is indicated by the onset of lactation or galactorrhea. Perhaps the clearest evidence for the secretion of prolactin *in vivo* by isolated pituitary tissue comes from work with pituitary autotransplants in the hypophysectomized rat (Everett, 1966). When this operation was carried out at oestrus, with fresh corpora lutea present in the ovaries, luteal function could be maintained almost indefinitely. However, prolactin secretion occurred whether or not the ovaries contained corpora lutea, as could be demonstrated by causing corpus luteum formation after an interval by giving gonadotrophin. Replacement of autotransplants of pituitary tissue beneath the median eminence led to a resumption of the secretion of FSH and LH by the graft, curtailment of the release of prolactin, and the return of cyclic ovarian function. And this despite the double insult of transfer of the pituitary gland, first to the kidney, and then back to the hypothalamus. The release of prolactin has also been demonstrated in intact females. In this case the grafts were homotransplants and the sexual rhythm of the hosts changed from a sequence of oestrous cycles to a series of pseudopregnancies. This effect (the Mühlbock-Boot effect) was not achieved by release of stored hormone from the grafts, for a series of pseudopregnancies could continue for more than a year. Male pituitary glands were as effective as those from females in this situation. Besides being continuous, there is evidence that the secretion of prolactin by transplanted pituitary tissue is greater than that of the normal cyclic rat. Pituitary grafts placed under the renal capsule secrete enough prolactin to initiate milk secretion in the oestrogen-primed rat or to maintain lactation in the hypophysectomized animal also given adrenocortico-trophic hormone. By itself, a transplanted pituitary located in a hypophysect-omized rat favours only mammary growth.

Lesions in the median eminence act like pituitary stalk section in promoting prolactin secretion, and in view of the probable inhibition exercised through the brain, it may be expected that lesions elsewhere in the hypothalamus may also favour pseudopregnancy or lactation. Prolonged periods of dioestrus, suggestive of pseudopregnancy, have been observed in rats with lesions in the neighbourhood of the paraventricular nuclei, and lactation occurred in male rats after placement of lesions in the tuberal area. Galactopoiesis ensued in oestrogen-treated female rabbits after damaging the basal posterior tuberal region, but in cats, mammary growth and lactation followed transection of the floor of the hypothalamus behind the optic chiasma. It is not yet clear whether the responses in cats and rabbits reflect the existence of different locations of controlling zones within the hypothalamus, or simply imply that the important area is diffuse. There does seem to be a species difference between cat and rabbit in the location of the oestrogen-sensitive area concerned with sexual behaviour.

Although both ovulation and pseudopregnancy commonly follow electrical stimulation of the hypothalamus of the rat, the processes are nevertheless separable. Ovulation which was not followed by pseudopregnancy has been caused by an irritative lesion placed in the medial preoptic area. Repetition of this procedure, together with stimulation of the tuberal region, induced pseudopregnancy, as did stimulation of the medial hypothalamus alone (Everett and Quinn, 1966). It may appear paradoxical that the release of prolactin is restrained by the brain and yet can follow excitation of the hypothalamus, but the stimulus must in some way remove the inhibition. Delayed pseudopregnancy can also occur after cervical stimulation on the first or second days of dioestrus, or after a variety of experimental interventions. Under these circumstances, the cycle in progress ends normally and a pseudopregnancy begins after the next ovulation (Everett, 1968). By some means, the effect of the brief stimulus must be retained for some days during one cycle so that luteal function can be extended during the next, and it may be that prolactin secretion is accelerated immediately but can exert no action in the absence of responsive corpora lutea. This would imply that the brief triggering stimulus, whether from coitus or manipulation of the cervix, must initiate a new pattern of central nervous activity that persists for some days, and in which the feedback action of progesterone from the corpora lutea is unlikely to be concerned.

Rather little is known about the role of extra-hypothalamic structures in the control of prolactin secretion (Tindal, 1967). Removal of the telencephalon in oestrogen-primed rabbits caused milk secretion, while more restricted lesions in the neocortex and olfactory bulbs did not do so. The amygdaloid region in the temporal lobes is known to be involved in the control of LH secretion and there is evidence for an influence on prolactin release. Bilateral lesions in the amygdaloid complex may cause lactogenesis in the pseudopregnant rabbit or the oestrogen-primed spayed animal, and minute bilateral implants of oestrogen act likewise.

Gonadal Hormones and Prolactin Secretion

The release of prolactin is enhanced by oestrogen, but it is not clear whether the steroid acts directly upon the hypophysis or through the hypothalamus. Large amounts of oestrogen raise the prolactin content of grafted pituitary tissue, but small implants of oestrogen made directly into the pars distalis of rabbits do not increase the concentration of prolactin, unlike similar implants in the posterior tuberal region of the hypothalamus. In rats, the direct application of a mixture of oestradiol and progesterone (1:10) to the pituitary gland prolonged the duration of pseudopregnancy. A similar mixture of oestradiol and cholesterol, or pure progesterone, was without effect (Francis and Malven, 1968). Oestradiol, thyroxine, and tri-iodothyronine added to cultures of pituitary tissue increase the amount of prolactin released into the medium (Meites, 1967), but it is difficult to apply these findings to the whole animal. However, testosterone, progesterone, cortisol, and corticosterone do not alter prolactin release from incubated pituitary glands.

Single doses of oestrogen injected into rats during prooestrus or oestrus may initiate a pseudopregnancy. Since this condition lasts for some 12 to 14 days, and prolactin is required throughout, a lasting change in hypothalamic function would seem to have been caused. Rather similar findings have been obtained with progesterone, which is considered to have a positive feedback action on prolactin secretion (Pasteels and Ectors, 1968), and it has been suggested that both steroids may enhance the sensitivity of a neural mechanism which then responds to stimuli which would otherwise be below threshold. The neural mechanism would then promote the release of prolactin (Everett, 1966).

If progesterone exerts a positive feedback action on prolactin secretion then a question arises as to the manner in which the release of hormone is terminated, for the corpora lutea of the rat may function for months and the secretion of prolactin by pituitary tissue can be long continued. No entirely satisfactory answer to this problem can be given but there is evidence that LH is luteolytic in the rat and rabbit (Rothchild, 1965). Since the corpora lutea are governed by other means in species such as the guinea-pig and sheep, this problem is not generally encountered.

Mammary gland development occurs during pregnancy, but lactation does not normally set in until after parturition. Because of species differences it is difficult to generalize about the factors initiating the release of prolactin on delivery, but these include a possible insufficiency of adrenal corticoids (since cortisol injections given to the pregnant rabbit can induce lactation), an antagonism between oestrogen, progesterone, and prolactin in their actions on the mammary gland, and neural stimuli from the genital tract at delivery and from the mammary glands with the suckling of the newborn.

Prolactin Secretion and the Release of FSH and LH

Studies of the way in which reserpine induces pseudopregnancy in the rat, and influences prolactin secretion in other species, have become integrated with recent anatomical studies on the hypothalamus and have led to the revival of an old concept: that the sympathetic system may be involved in the control of gonadotrophin secretion. The original suggestion was based upon the blockade of ovulation in the rat by a range of drugs which interfered with transmission at the synapses in the sympathetic system, and with the induction of ovulation by sympathomimetic agents. It was suggested, perhaps prematurely, that the neurohumoral transmitter from the hypothalamus to the pituitary gland for gonadotrophin release and ovulation was adrenergic in character, but this was disproved (Donovan and Harris, 1955). Subsequently, interest in the relationship between the sympathetic system and gonadotrophin secretion tended to wane until it was realized that reserpine, which readily induces pseudopregnancy in the rat, reduces the amount of noradrenaline present in the brain and that other drugs which similarly influence the catecholamine content of the brain also cause pseudopregnancy. Substances which deplete only the peripheral stores of noradrenaline are not active in this way. It is not necessary to give large doses of reserpine, which, besides causing pseudopregnancy, has marked sedative effects and depresses food and water intake; the local implantation of minute amounts of the drug into the median eminence has proved equally effective in the rat, and also depleted the monoaminergic material in its immediate neighbourhood (van Maanen and Smelik, 1968). The concurrent systemic administration of a monoamine oxidase inhibitor (iproniazid) prevented pseudopregnancy, but these workers incline to the view that the active system is dopaminergic. Alongside the induction of pseudopregnancy, the ovulation expected in immature rats after the administration of pregnant mare serum has also been prevented by reserpine, provided treatment was initiated before the critical period for LH release. Cyclic ovulation in the rat and hamster has been blocked by treatment with reserpine or with α-methyltyrosine.

Since the secretion of FSH and LH is often depressed in circumstances when that of prolactin is raised, it is feasible that a reciprocal relationship exists between the secretion of FSH and LH on the one hand, and prolactin on the other. Prolactin secretion is favoured, and that of FSH and LH depressed, by chronic oestrogen treatment or by the administration of tranquillizing drugs. A parallel situation is found after pituitary stalk section or transplantation. The stimulus of suckling also tends to suppress gonadotrophin secretion and to raise that of prolactin. However, this relationship is not immutable. A suckling stimulus sufficient to establish a pseudopregnancy does not prevent the hypersecretion of gonadotrophin that occurs after castration, and there is evidence that LH secretion sufficient for ovulation can still occur in rats in which prolactin secretion has been promoted (Everett, 1966). The females of a number of species ovulate immediately after delivery, when there is a concurrent release of prolactin and LH, and it has been shown that different hypothalamic areas are concerned with prolactin release (initiation of pseudopregnancy in the rat) and ovulation. Further, the releasing factors for LH and prolactin are separable.

Prolactin Inhibiting Factor

At first sight, the fact that pituitary tissue divorced from the hypothalamus secretes prolactin in relative abundance may be taken to indicate that there is no hypothalamic releasing factor for this hormone. But the correct conclusion is that the hypothalamic factor is inhibitory in nature (McCann and Porter, 1969). Pituitary tissue cultured *in vitro*, secretes substantial quantities of prolactin into the medium, and the addition of an extract of median eminence depresses the release of the hormone without apparently damaging the pituitary tissue. Extracts of cerebral cortex have been frequently applied to pituitary tissue for control purposes, for they do not alter the output of any trophic hormone. However, the hypothalamus is a storehouse of pharmacologically active substances and it is always advisable to examine a range of naturally occurring agents. Oxytocin, vasopressin, adrenaline, acetylcholine, serotonin, histamine, and bradykinin have been tested and do not inhibit the release of prolactin from cultures of pituitary tissue. Suckling elicits both the release of oxytocin and of prolactin from the hypophysis and it has been suggested that oxytocin may act as the neurohumoral agent for prolactin secretion. This view has lost popularity for, among other evidence, oxytocin does not cause release of prolactin when applied directly to the pars distalis. Suckling, as well as treatment with reserpine, chlorpromazine, progesterone, or oestradiol, reduced the hypothalamic content of the prolactin inhibiting factor (PIF) and these results agree with the idea that prolactin release is stimulated by depression of hypothalamic PIF production. Hypothalamic, but not cerebral cortical, extracts could block the nursing-induced decline in pituitary prolactin, and pig hypothalamic extracts have been reported to block the depletion of pituitary prolactin that follows cervical stimulation in the rat. As with other hypothalamic releasing factors, there is a dose-response relationship between the amount of extract added to a pituitary culture and the amount of prolactin released. In the case of prolactin, the quantity released is inversely proportional to the log of the dose of extract, for glands from either male or female rats. Under experimental conditions, exogenous prolactin can inhibit the endogenous secretion of prolactin by the hypophysis. The pituitary content of prolactin was depressed in rats carrying tumours of pituitary tissue that secreted prolactin, or grafts of pituitary under the capsule of the kidney. Implants of prolactin made into the median eminence of adult female rats increased the amount of prolactin inhibiting factor in the hypothalamus, depressed the concentration of prolactin in the hypophysis, and limited luteal function as shown by a resultant shortening of the length of pseudo-pregnancy (Chen, Voogt, and Meites, 1968). In immature rats, prolactin implants made into the median eminence have been found to enhance follicular growth and uterine weight, probably by causing the release of FSH (Voogt, Clemens, and Meites, 1969). The mechanism involved remains unclear, but this effect provides a further example of the reciprocal relationship between the output of prolactin and that of the other gonadotrophins (p. 130).

12

The Nervous System and the Function of the Pars Intermedia

Despite the existence of a large body of knowledge concerning the activities of the pars intermedia in lower vertebrates, where it is concerned with colour adaptation, little is known about its function. Much of the following is based upon the reviews of Harris (1966), Lerner (1966), Wingstrand (1966), Etkin (1967), and Novales (1967).

Lying between the pars distalis and pars nervosa of the hypophysis, the intermedia is a variable structure, being composed of a few layers of cells in some mammals, such as the rat and dog, while in others it may be quite large. In yet other cases (whale, elephant, armadillo, and man) the pars intermedia is absent or rudimentary and it has been suggested that an intermingling of pars intermedia cells with those of the pars distalis may have occurred.

The pars intermedia is innervated from the hypothalamus and the nerve supply may include neurosecretory fibres. Although the source of the fibres is not known exactly their overall effect is inhibitory, as indicated by the physiological evidence of skin darkening in amphibia and the fact that hypertrophy of the pars intermedia appears to occur after division of the pituitary stalk, or transplantation of the hypophysis. Compared with the pars distalis or neuralis, in most species the intermedia is poorly vascularized and in general the capillary beds of the intermedia and neural lobes receive a common arterial supply; the intermedia is not primarily fed by the portal vessels.

The hormones of the pars intermedia have been characterized and their structures are known. There are two forms, α-melanocyte stimulating hormone and β-melanocyte stimulating hormone, with the former being a straight chain polypeptide with 13 amino acids which is present in the glands of all mammals and has been synthesized. Secretion of β-melanocyte stimulating hormone occurs (it has been detected in human plasma by immunoassay procedures), but the α-form has not been detected in the circulation. There are minor differences in the structure of the β-melanocyte stimulating hormones between species depending upon the sequence of amino acids at the ends of the molecule. All the melanocyte stimulating hormones (α and β alike) contain a sequence of 7 amino acids (methionine, glutamine, histidine, phenylalanine, arginine, tryptophane, and glycine) and this sequence is also to be found in the molecule of ACTH. The melanocyte stimulating activity displayed by ACTH may be due to this similarity in structure.

Melanocyte stimulating hormones disperse the pigment in the melanophores of amphibian skin and may, with prolonged treatment, cause an increase in the total quantity of pigment. The high activity of α-melanocyte stimulating hormone is indicated by the fact that the injection of 0·0025 µg (50 units) into a frog causes rapid darkening. On a molar basis, the β-form has about half the potency of α-melanocyte stimulating hormone on frog skin, and ACTH has about one-thirtieth the activity of the α-hormone (Lerner, 1966). Serial injections of MSH or of ACTH into men or women can bring about a darkening of the skin after a week or more, and a bronzing of the skin is occasionally seen in cases of adrenal insufficiency when the melanocyte stimulating activity of the blood becomes abnormally high. The darkening of mammalian skin is not to be equated with that of amphibia since the pigment of mammalian melanocytes, unlike that of amphibian melanophores, is not mobile. In mammals the absolute amount of melanin in the skin is increased.

The physiological function of melanocyte stimulating hormone in mammals is not known. It might be concerned with the mobilization of fat, for the administration of the hormone raised the concentration of fatty acids in the blood of experimental animals. However, other peptides have this effect and the significance of the response is obscure. The same may also be said of the increase in the amplitude of monosynaptic potentials in the spinal cord brought about by β-melanocyte stimulating hormone. Thus the mammalian intermediate lobe provides an excellent example of a gland secreting a hormone of unknown function. There is little doubt that secretion of intermediate lobe hormone does occur, for melanocyte stimulating activity is increased in the blood and urine of individuals suffering from adrenocortical insufficiency, while the activity of body fluids falls after hypophysectomy or treatment with steroid hormones. It may be argued that much of the melanocyte stimulating activity is due to the presence of large amounts of ACTH, but this is unlikely, particularly in view of the results obtained with a sensitive immunoassay for the β-hormone.

The hypothalamus probably inhibits the secretion of intermediate lobe hormones. In amphibia darkening of the skin ensues after stalk section and a number of experimenters have remarked on the enlargement of the pars intermedia after this operation in mammals. Unfortunately, there is little direct evidence of an increased secretion of hormone from the mammalian gland and most conclusions derived from experimental work are based upon changes in the content of hormone. Howe and Thody (1969) have found that lesions in the neighbourhood of the paraventricular nuclei of the hypothalamus caused an increase in the MSH content of the intermediate lobe. Sham operations, which involved puncture of this area, also elevated the MSH content but this was of short duration, while that produced by the lesions lasted. Enlargement of the pars intermedia also occurred in the lesioned animals and there were histological signs of hyperfunction of the cells, as well as an increased incidence of mitosis. However, the lesioned animals were no darker than controls and no indications of the action of any hormone released were detected. Reserpine depresses the pituitary content of melanocyte stimulating activity in toads and rats and since the drug causes darkening of the skin in toads it would seem to cause hormone release. This action of reserpine parallels that of the drug on prolactin secretion (p. 126) and is in accord with the existence of an inhibitory mechanism.

It is curious that there seem to be two neurohumoral factors extractable from the hypothalamus which influence the release of melanocyte stimulating hormone. One promotes secretion, as with the pars distalis releasing factors, and the other blocks the discharge, being called the release-inhibiting factor (Tomatis and Taleisnik, 1968; Schally, Arimura, Bowers, Kastin, Sawano, and Redding, 1968). From 5000 cattle hypothalami, weighing 2500 grams, Schally and Kastin (1966) recovered 38 mg of purified melanocyte stimulating hormone-inhibiting factor, of which 0·06 μg was able to elevate the MSH content of the pituitaries of normal rats. It is difficult to reconcile this information with the evidence that the control of the intermediate lobe is exercised by nerve fibres, and not a vascular pathway, unless the factors mentioned above are released at the nerve terminals. In toads, interruption of the posterior lobe nerves interfered with the neural control of intermediate lobe function for months, whereas regeneration of divided vascular pathways, as with the portal vessels, required much less time (Jorgensen and Larsen, 1963; Etkin, 1967).

13

Rhythms in Neuroendocrinology

Rhythmic fluctuations in bodily activities and functions have been known for many years and have provided the basis of an extensive literature. In 1797, Christopher William Hufeland wrote in his *The Art of Prolonging Life* (2nd Edition, London):

> 'The period of 24 hours formed by the regular revolutions of our earth, in which all its inhabitants partake, is particularly distinguished in the physical economy of man. This regular period is apparent in all diseases; and all the other small periods, so wonderful in our physical history, are by it in reality determined.'

Since then, cycles in body temperature, behaviour patterns, locomotor activity, urine excretion, the concentration of blood constituents, endocrine function, and reproductive pattern have been described in detail, without, it must be said, greatly advancing understanding of the mechanisms involved.

The symptoms of various illnesses become manifest with great regularity in some patients. Instances are known where the body temperature rose and fell three degrees every 12 hours, while in one case, the body temperature reached a peak of 40°C just after midnight every day. This patient was studied as a boy of seven and a half, and still showed the same pattern 13 years later. A regular diurnal rhythm in the timing of fits in epileptic patients has been observed. Rhythms of susceptibility to noxious influences have been plotted. In mice kept under a 12-hour light, 12-hour dark sequence, constant doses of acetylcholine are much more toxic when given 2 hours after the onset of the daily dark period than when given 2 hours after the onset of the light period. The peak mortality after toxic doses of ethanol in mice, like that to audiogenic seizures in certain strains, also occurs 2 hours after the onset of the daily dark period, and the susceptibility rhythm can be shifted by reversing the light-dark pattern.

A diurnal rhythmicity in the secretion of hormone is shown by a number of endocrine glands, with that of the adrenals being most closely studied. In rats, the blood levels of ACTH and an adrenal corticoid, corticosterone, vary in phase, with peak levels being measured just before the onset of darkness and the lowest values at dawn. A shift of 12 hours in the lighting sequence alters the peak of corticosterone secretion by a similar period, so that the rhythm is evidently controlled through environmental lighting. Exposure to constant light depresses the secretion of corticosterone and abolishes the peak expected at dusk; in adrenalectomized animals the rhythmic secretion of ACTH is also suppressed (Cheifetz, Gaffud, and Dingman, 1968). Further, the rhythm is abolished in animals in which the hypothalamus has been isolated from the rest of the brain. Variation in the secretion of ACTH is not necessarily the sole factor in producing the cycle of adrenal activity, for there is also a diurnal rhythm in the response or reactivity of the adrenal gland to its trophic hormone, with the greatest sensitivity being found just before dawn.

In man, the highest levels of adrenal corticoids occur between 6 a.m. and 8 a.m., with a gradual decline toward minimal levels at 6 p.m. and midnight. The changing levels are not due to changes in metabolic rate or to diurnal rhythms of responsiveness of the adrenals to ACTH, since they are not seen in patients suffering from adrenal cortical insufficiency given constant doses of cortisone, or in subjects infused with a constant amount of ACTH throughout the day and night. The best available evidence indicates that rhythmic changes in ACTH secretion occur in which the feedback action of adrenal hormones can play only a minor part. This is because the diurnal rhythm in pituitary-adrenal activity is coordinated with the sleep pattern of the individual. When this is altered, so adrenal function changes. Subjects who have become accustomed to two cycles of sleeping and waking every 24 hours display two cycles of adrenal activity during this period. Several attempts have been made to explain the circadian rhythm in adrenal function. Mills (1966) suggests that a variety of periodic influences impinges upon the central nervous system and perhaps upon the reticular formation, with the most important being social contact with other people whose activity and sleep habits follow a circadian pattern. Excitation resulting from such stimuli is then transmitted to the hypothalamus and results in varying levels of release of the neurohumoral agent for the discharge of ACTH. The difficulty with this hypothesis is that under other circumstances stimuli leading to the release of ACTH act very quickly indeed, whereas peak levels of adrenal activity can occur in association with the quiescence of dreamless sleep.

The mechanisms controlling the circadian secretion of ACTH can be differentiated from those involved in ACTH release as a response to stress. Anterior hypothalamic lesions in cats have been found to abolish the circadian periodicity in plasma 17-hydroxycorticoid levels without causing loss of the normal rise seen after exposure to stressful noise; and the implantation of cortisol into the hippocampus or midbrain of rats disrupted the diurnal pattern of adrenal activity (Slusher, 1964, 1966). Peaks in adrenal cortical activity in cats occur between 8 p.m. and 4 a.m. Treatment with atropine at 6 p.m., just before the expected rise, prevents the anticipated change in plasma 17-hydroxycorticoids, while provision of the drug at other times of the day does not alter the rhythm. A similar effect has been obtained with a short-acting barbiturate. Further, treatment with atropine does not interfere with the adrenal reaction to insulin hypoglycaemia, so implying a rather specific action on the circadian mechanism (Krieger, Silverberg, Rizzo, and Krieger, 1968).

Rhythmic patterns in the function of several endocrine glands can occur independently of one another, although it is necessary to bear in mind that different hormones act on target tissues at different rates. This means that the results observed, although separated in time, could be due to the simultaneous release of several hormones, and underlines the value of measurement of the blood levels of the hormones themselves. The release of ACTH from the pituitary gland of the rat, which is at its height at dusk, occurs at a different time from that of luteinizing hormone, which takes place at mid-afternoon. Alteration of the secretion of gonadotrophin, as occurs with spaying, does not disturb the diurnal corticoid secretion pattern in the rat. Relatively little change in the thyrotrophic hormone content of the rat hypophysis occurs during the day although the lowest level is found coincidentally with the peak in plasma concentration, at 3 p.m. (Singh, Panda, Anderson, and Turner, 1967). Thus the circadian rhythm in TSH secretion appears to be approximately in phase with that of ACTH.

Under conditions of controlled illumination a diurnal rhythm in the timing of ovulation in the rat is demonstrable. This has emerged very clearly from the work of Everett and his colleagues, who studied various aspects of the blockade of ovulation with pharmacological agents (Everett, 1964). When rats are illuminated from 5 a.m. to 7 p.m. daily, a surge of gonadotrophin secretion from the pars distalis occurs between 2 p.m. and 4 p.m. on the day of prooestrus, with ovulation taking place about 10 hours later. Treatment with barbiturates, morphine, reserpine, chlorpromazine, anticholinergic, or anti-adrenergic drugs over this period will prevent ovulation, but not when the drugs are administered after 4 p.m., for by that time sufficient hormone has been released to cause follicular rupture. In the absence of further treatment, ovulation ensues 24 hours later than expected, but can be delayed for an additional day by repetition of the blockade over the critical period between 2 p.m. and 4 p.m. A critical time period for ovulation is also demonstrable in immature rats treated with pregnant mare serum gonadotrophin. Experiments involving hypophysectomy or the administration of barbiturates show that, as in the adult, the release of gonadotrophin for ovulation occurs between 2 p.m. and 4 p.m., and that blocking agents must be given before 2 p.m. to

137

inhibit follicular rupture. Since it has been recently found that there is no marked fall in the pituitary content of ovulating hormone by the end of the critical period, the release of gonadotrophin may continue for some undetermined period (McCormack and Meyer, 1968). It seems that the daily waxing and waning in the secretion of gonadotrophin in the rat is not dependent upon the presence of gonadal steroids, for cyclic fluctuations in the plasma concentration of luteinizing hormone have been measured in rats studied five months after ovariectomy, although fluctuation in the disappearance rate of the hormone from the plasma could possibly account for this result (Lawton and Schwartz, 1968). The timing of the critical period for ovulation in the rat (which may vary somewhat from strain to strain) can be altered by moving the period of illumination around the clock, although the shift to a new setting is a gradual one and takes a fortnight to complete. A diurnal cycle in the amounts of FSH and LH stored in the pituitary gland of the male rat has also been described, with the content of each hormone being roughly twice as high in the afternoon than in the morning (Martini, Fraschini, and Motta, 1968). Much less information is available for species other than the rat and mouse. Diurnal variations in the onset of oestrus and in the success rate of artificial insemination of cattle have been reported.

Seasonal Rhythms

Annual rhythms in endocrine function are mostly concerned with the timing of breeding and of reproductive activity, although they may also be necessary to promote survival during the difficult conditions of winter. The deposition of fat and the change of coat provide familiar examples of the latter. It is almost axiomatic that the changes in endocrine function must be initiated long before winter, or before the breeding season, so that a thick coat has been grown, or full gonadal development achieved, at the proper time. Most commonly, day length provides the cue for the timing mechanism because of the high degree of reliability of this factor, and photoperiodic mechanisms have been demonstrated in almost all classes of vertebrates. Evidence is more widely available for the control of ovarian cycles and oestrus in mammals than for the control of testicular activity, but this simply reflects the lesser attention given to the male because of the difficulty of tracing testis function over long periods, although fluctuations in the constituents of semen have been followed in domestic animals. In birds, by contrast, the male has been more widely used than the female.

Under constant conditions, annual cycles may continue, though they are no longer geared to environmental changes. Ferrets have been kept under short-day conditions from birth, from the beginning of the year, or placed in darkness, with little or no delay in most instances of the onset of oestrus. The results of work with blinded animals also indicate that oestrus recurs at irregular intervals. Exposure to continuous illumination also seems to disrupt the endogenous rhythm, and is less effective than long days in promoting sexual development (Donovan, 1966).

At present, the way in which the external stimulus of light is employed as a cue is not understood. Nor is it known how in some species, such as ferrets, horses, and donkeys, significance is attached to long days, whereas in others, such as sheep and goats, it is the short days that are important. Several variables attached to the length of day could be utilized by the nervous system:

the gradual change in the period of illumination, the number of hours of daylight, or the ratio of light to dark, but the first two do not seem to be important. Ferrets will come into heat precociously when transferred in the autumn from short to long days without any transition period between the two regimes. Although a ratio of light to darkness of 2:1 is highly effective in promoting sexual development in this species, premature oestrus can be precipitated by exposure to an hour of light at midnight, and a schedule of 2 hours light, 10 hours dark, 2 hours light, and 10 hours dark is also effective. Ortavant, Mauleon, and Thibault (1964) suggested that the light photoperiod facilitates the release of gonadotrophic hormones, whereas the dark period would vary from species to species and the light periods must be sufficiently long or close together to elicit a response. When the days are too short, the release of gonadotrophic hormone wanes, but flashes during the dark period can result in sexual activity. On the other hand, the prolongation of long light photoperiods brings about a decrease in the production of pituitary hormone due to the development of fatigue in the neural processes involved. While this concept may have the advantage of not requiring the mechanism for 'short day' animals to differ from that of 'long day' animals, it is not readily amenable to experimental test.

Interaction between seasonal changes in daylength and endogenous diurnal rhythms may be significant. Evidence for a daily variation in the sensitivity of the response to photic stimuli has come from work on birds (Farner, 1965; Wolfson, 1966) where two phases in the regulation of the annual gonadal cycle can be distinguished. A preparatory phase occurs in the late summer and autumn in response to short days, and is required for the development of sensitivity to long days on the part of the neural mechanism controlling gonadotrophin secretion. In its absence (as in constant exposure to long days) gonadal development in the spring does not occur. After the long preparatory phase is over, the bird becomes responsive to long days and so a phase of sensitivity to light, or a progressive phase, ensues. It is during the progressive phase that a period of sensitivity at particular times of the day becomes apparent and seems to be linked with a period of motor activity. The significance of the correlation between the duration of motor activity and the response of the gonads is a matter for speculation. Wolfson (1966) suggests that the duration of activity could be a reflection of the time-measuring process in the brain, which uses an endogenous clock and the duration of the light period to measure the length of day, and which in turn governs both the secretion of gonadotrophin and the duration of motor activity, or that the hypothalamus is directly involved and controls both functions simultaneously. Ten-minute periods of illumination applied during the free-running rhythm of activity of flying squirrels kept in darkness, provided a biphasic curve in that near the time of the spontaneous onset of locomotor activity the stimulus delayed the subsequent onset of activity, but when given some 8 hours later, the onset of activity was advanced. Similar findings have been obtained in hamsters kept in a constant dark environment or exposed to a 12-hour day (De Coursey, 1964).

Circadian Rhythms

Reference has been made on several occasions to diurnal rhythms in endocrine function, but since the term diurnal refers particularly to the day, this usage can be misleading. Nocturnal animals can have diurnal rhythms too. It was to overcome this difficulty that Halberg (Halberg, Halberg, Barnum, and Bittner, 1959) introduced the term 'circadian' for rhythms which have a period length of about one day; the patterns of adrenal and gonadal activity seen in the course of 24 hours, and referred to earlier, come under this heading.

Many studies of circadian rhythms have utilized the regular changes in activity and rest exhibited by the species under examination. This parameter can be followed quite readily for long periods without disturbing the subject, or the need for collecting samples of body fluids, and has provided information of neuroendocrine interest. Rhythms of running activity in the female rat can be traced which are closely linked with the various phases of reproduction. They are not seen in the male. The female rat shows a peak of running activity every 4th or 5th day which coincides with oestrus, and there is a fall in running activity during pseudopregnancy. In women, too, the number of miles walked every day shows cyclic fluctuations and reaches a maximum near ovulation at the midpoint of the menstrual cycle.

Under constant conditions, such as darkness, there may be a constant drift in the rhythm of activity away from synchrony with an external clock because the periodicity is not exactly 24 hours. But in individual animals a circadian rhythm may remain constant. Flying squirrels kept in darkness show a circadian pattern of activity which may deviate ± 2 minutes from a mean value. Many records of activity of individual rats have been described by Richter (1965), and these show that individual rats can keep time under constant, or free-running, conditions with astonishing precision. One animal began running precisely 26 minutes earlier on successive days. Others started running 36 or 38 minutes earlier each day. These observations are taken to imply that the biological clock is of endogenous origin, for such timing is divorced from external cues. An underlying 24-hour rhythm of activity has been found to persist in rats reared for twenty-five generations in continuous light. Confinement of human subjects in a sound-proof bunker without time cues did not abolish the diurnal rhythm present in a variety of bodily functions. In one study (Aschoff, 1965), there was a clear cycle of sleep and wakefulness which was reflected in the rhythms of body temperature and urinary excretion. The maxima of the functions studied were not always exactly in phase with each other but all showed a periodicity of about 25·0 hours. In another subject, the sleep-wakefulness rhythm had a mean period of about 25·9 hours, and as a result of this, after 18 physiological days of confinement, his phase of activity was dissociated from that customary under normal conditions. On the last day of the experiment he woke up at 8 p.m. and then immediately readjusted to the correct phase relationship with normal time by being active for more than 30 hours.

Experiments of a similar, but more drastic, nature have been performed by Lindsley and his colleagues (1964) who reared six infant monkeys in isolation and darkness for periods up to 6 years, except for diffuse unpatterned light stimulation for 1 hour each day to prevent retinal degeneration. Measurements of spontaneous activity taken over 12 months showed 12- to 16-hour diurnal activity periods, with 12- to 8-hour periods of relative inactivity. The activity cycles were not affected by shifting feeding schedules but were anchored to, and moved with, light stimulation periods. From 3 to 5 weeks were required on a given schedule to stabilize activity periods after a shift in the light period with respect to solar day, and splitting the light periods into 2 half hours separated by 11 hours tended to produce a bimodal distribution of activity, each anchored to light stimulation.

With exposure to regular alternation of night and day, the circadian rhythm rapidly becomes locked on to the external environmental change. When the lighting conditions are held constant, as with continuous darkness or continuous illumination, other cues such as temperature changes or a regular schedule of feeding may be utilized. The cues or entraining agents are sometimes called 'Zeitgebers'. It is difficult to entrain animals to day lengths of less than about 16 hours. This is illustrated by work on domestic mice where activity cycles could be entrained to day lengths of 21–27 hours. Outside these limits, the mice reverted to a 24-hour cycle of their own. Hamsters exposed to a 6-hour day of 2 hours light and 4 dark, extracted a 24-hour rhythm from the pattern by responding only to every 4th cycle; and if a 4-hour day was imposed then the animals responded only to every 6th cycle.

An internal drive toward rhythmicity seems so predominant that a cyclic pattern generally persists under constant conditions. It is experimentally difficult to eliminate all possible rhythmic changes, for fluctuation in barometric pressure, humidity, and cosmic radiation have all to be taken into account. In part, the operation of an endogenous clock can be tested for by transhipping the experimental animals elsewhere to another part of the globe, where conditions are different, and observing whether the existing rhythm persists. Efforts have been made to stop the clock experimentally. Success has been achieved in bees trained to visit a food dish at a certain time of day, when it was found that individuals cooled for 5 hours showed a shift in rhythm of a similar period. However, this procedure is not easily applied to mammals, although Richter (1965) was able to arrest the rhythm of activity in blinded rats by the induction of hypothermia, in which the body temperature fell to 7°C or below, and both respiration and the heart beat ceased for an hour or more. Elevation or depression of the metabolic rate, removal of the adrenal glands, gonads, thyroid, pineal, or pituitary glands did not alter the rhythm, while electroshock, or the production of convulsions, or stupor were equally ineffective.

Circadian rhythms appear to be remarkably independent of changes in temperature. Normally, metabolic processes go on more rapidly with rise of temperature so that a circadian rhythm would be expected to accelerate to a degree which can be calculated. This does not occur, and implies that circadian rhythms are not merely reflections of metabolic changes. However, the existence of a metabolic basis for a circadian rhythm is implied by study of the activity pattern of deer mice, *Peromyscus leucopus noveboracensis,* given deuterium oxide in the drinking water (Suter and Rawson, 1968). The period length increased when deuterium oxide was supplied, and was shortened when the heavy water was withdrawn, so that a linear relationship between period length and deuterium oxide concentration could be established. Hamsters and bats follow a 24-hour rhythm of awakening from hibernation, and this is despite a prolonged period of low body temperature. Circadian rhythms in the brain temperature and EEG of the septum of the squirrel have been described, although the EEG of the olfactory bulb shows only a weak rhythm. Such findings lead to the suspicion that circadian rhythms are not common to all brain areas and that oscillations of higher and lower frequency exist. Under the special conditions of hibernation, oscillations of electrical activity in the amygdala of the squirrel with a periodicity of 4 to 9 hours have been observed. When the amplitude of the oscillation becomes critically large, arousal from hibernation occurs.

No satisfactory explanation of the basis of circadian rhythms is yet available, although theories involving temperature-sensitive metabolic rhythms, the accumulation of toxic depressants, and the elaboration and exhaustion of reserve products can be discarded. Some have been impressed by the evidence that many different kinds of organisms exhibit fluctuations in metabolic rate that can be correlated with specific, unpredictable, and local parameters of barometric pressure, air temperature, or background radiation, and consider that the timing mechanism is of exogenous origin (Webb and Brown, 1959). The existence of three functional oscillators has been postulated, with the first being the externally imposed fluctuations providing reference frequencies, the second being the monitoring system which has some capacity of its own to oscillate, and the third system being the indicator or observed process or processes driven by the second oscillator. Because of present difficulties in comprehending the means by which subtle geophysical changes may be detected by an individual, this concept has not won wide acceptance. It is always necessary to exercise care when extracting rhythmic changes from a time series of data, as is illustrated by the cautionary tale of the study of the biological clock of the unicorn. In this case, the values for metabolic activity were extracted from a table of random numbers. At first the results seemed to suggest underlying rhythms, but no pattern was clearly apparent until it was realized that in summer at 40° north latitude, the hour of the rise of the moon may be retarded by approximately one hour each night. When this rhythm was allowed for and eliminated, a daily pattern of metabolic rate change emerged. As Cole (1957) remarks, the rhythm 'could easily be shown to be highly correlated with environmental fluctuations, but the nature of the material employed in this experiment seems to preclude any such causal relationships'. It may be that, as Harker (1958) has suggested, there is a basic 24-hour metabolic rhythm present in the cells of all animals, that this is inherited, that it continues unchanged at a cellular level even when it is not evident in behavioural or major physiological changes, and that it may be masked by a predominant influence of the environment. The phases of the circadian rhythms of different cells may be set by different environmental factors, and so any cell, or group of cells, may constitute a 'physiological clock' which may in turn affect the phases of the rhythms of other groups of cells. Thus an animal may have a number of 'clocks', all in operation together. Most work upon the circadian rhythms of isolated cells has been done with invertebrates, but is of some relevance. Cockroaches show a diurnal cycle of activity which can be shown to originate in the suboesophageal ganglion, for when the suboesophageal ganglion of an insect with a particular circadian rhythm is transplanted into an arrhythmic host, the host animal takes up the rhythm of the donor. Corresponding information, but of a different nature, comes from work on *Aphysia californica,* the sea-hare. Strumwasser (1967) recorded from a single cell in the isolated abdominal ganglion of this species continuously for up to 48 hours. When the animal had been kept under constant light for 1 week prior to the experiment, the long recording periods revealed a free-running circadian rhythm in the long-term firing rate of the 'parabolic burster'. But if the animal had been exposed to alternating 12-hour periods of light and dark for several days before the experiment, then the 'parabolic burster' showed a marked increase in discharge rate, with the peak occurring close to the time when the switch from darkness to light would have been expected. After changes in the dark-light rhythm, the peak increase in the firing rate of the cell shifted appropriately. Heat pulses applied to the whole ganglion, or the intra-cellular injection of actinomycin D during the 10-hour period prior to the expected peak in firing rate, caused the peak to occur earlier, and Strumwasser suggested that these procedures caused a premature release of nuclear messenger RNA that initiated production of a cytoplasmic substance that depolarized the neurone membrane, thus causing the peak in spike activity.

Under constant illumination the free-running rhythm is influenced by light intensity. This principle has come to be known as Aschoff's law, which states that with increasing intensities of light the free-running period is shortened in diurnal animals and lengthened in nocturnal animals. Circadian rhythms in men studied in isolation are also affected by the level of illumination. With lighting of 40 lux an activity cycle of about 25·0 hours was recorded for one subject and this was shortened by 0·7–1 hour on increasing the intensity to 200 lux (Aschoff, 1965). When considered in relation to the general trend toward more intense lighting in modern society, this phenomenon may account for the acceleration of several bodily rhythms in man, of which the earlier occurrence of puberty may be one.

In much of the work concerning the influence of light upon endocrine function it is presumed that the effects of optic stimulation are transmitted to the hypothalamus to alter the output of the appropriate releasing factors. There is little doubt that the retina normally acts as a receptor in rats and ferrets but the pathway employed between the optic nerve and hypothalamus remains unknown, although light-evoked electrical responses in the hypothalamus are readily recorded. Direct connections between the retina and hypothalamus have been described but are not generally accepted, while connections running in the suprachiasmal region of the rat hypothalamus have been suggested (Critchlow, 1963; Donovan, 1966). Lesions in the medial forebrain bundle can block the response to light and are considered to interrupt an inferior accessory optic tract running to the midbrain (p. 151).

In insects, the release of the metamorphosing hormone ecdysone is triggered by a humoral agent released from certain brain cells of the dormant pupa. Release of the hormone is promoted by light which is absorbed by a pigment in the brain cells which is possibly of the porphyrin type. Porphyrins also occur in the mammalian nervous system and Klüver (1944) observed that the fluorescence spectra of the cranial nerves exhibited striking differences in the distribution of a pigment showing an emission band at about 625 mμ. It was present in the optic nerve but absent from the oculomotor nerve; present in the olfactory tract but lacking in the olfactory bulb. Klüver remarked that 'it seems particularly significant that the optic nerve should contain one of the most remarkable photodynamic substances ever discovered'. The pigment appears to be sensitive to light, because upon illumination of brain tissue the 625 mμ emission band rapidly disappears, so that the possibility arises that it could mediate a response to photic stimuli. Even in large mammals light can penetrate the skull, as shown by appropriate dissection of the head or by experiments in which a small light-sensitive photo voltaic cell was implanted into the brain of ewes, dogs, a rabbit, and a rat. It was established that light could reach the hypothalamus. In the duck, projection of light directly on to the hypothalamus through a quartz rod, after removal of the eyes and orbital tissues, caused genital stimulation (Benoit, 1964), and in blinded rats, light directed to the hypothalamus through fibres implanted under stereotaxic guidance, increased the percentage of cornified cells in the vaginal smear when compared to blinded controls.

Sexual Differentiation of Endocrine Cycles

Cyclic patterns of function are much more in evidence in the female than in the male. Female mammals show an oestrous or menstrual rhythm which is associated with the occurrence of cyclic patterns of behaviour; male gonadal function and behaviour is relatively constant and predictable. In the female, the occurrence of cycles in gonadotrophin secretion can be traced back into infancy and for the rat it is now clear that a sexual differentiation of the brain occurs around the time of birth. Removal of the testes allows retention of cyclic gonadotrophin secretion by the hypophysis of the male, whereas treatment of the female with androgen causes loss of cyclic ovarian function. There is a critical period from about the 18th or 19th days of intrauterine life to about 10 days postnatally during which the opposite pattern of sexual differentiation to that expected from the genetic sex can be imposed by hormone treatment. Outside these limits, experimental intervention is much less effective. It is unlikely that androgen acts on the brain over the whole of this period, for steroid hormones which can alter the process are effective with a single injection and it is unlikely that the steroid persists in the circulation for more than a few days. It might be better to refer to the critical period as a phase of sensitivity.

The female pattern of sexual function may be regarded as the basic or fundamental state, for it is this that is modified by androgen. The administration of testosterone to neonatal female rats abolishes the oestrous cycles expected to set in upon sexual maturation, and the removal of the testes of the newborn male allows retention of a cyclic pattern of gonadotrophin secretion, as shown later by the implantation of ovarian tissue (Harris, 1964). Put another way, androgen present about the time of birth in the rat stops the cyclic mechanism which elicits a rhythmic secretion of gonadotrophic hormones. Transplantation experiments have shown that the ovaries and pituitary glands of androgen-treated females retain their normal potentialities and indicate that hypothalamic function is disturbed. In other species with a longer gestation period the critical period for the sexual differentiation of the brain falls during foetal life and is much less readily investigated. Treatment of female guinea-pigs with androgen between days 30 and 65 of foetal life suppresses female behaviour and exaggerates the masculine features.

Damage by androgen to the mechanisms controlling the secretion of gonadotrophin can be detected early in infancy because of the occurrence of abnormal concentrations of gonadotrophin in the pituitary gland. However, after the vagina opens, oestrous cycles may occur for a variable period of time before persistent vaginal oestrus sets in, with the length of the interval of normal ovarian function being related to the level of androgen treatment. After large amounts of testosterone, the interval is very short before ovulation fails and the ovaries become purely follicular. The changes in brain function underlying the development of the syndrome are not known, but since the gonadal changes are closely similar to those seen after the placement of lesions in the anterior hypothalamus, it is reasonable to suppose that the mechanisms controlling ovulation have suffered damage. One view is that neonatal treatment with androgen desensitizes both neural and non-neural structures to the action of oestrogen, for the uterus of treated animals is less responsive to oestrogen than that of appropriate control females. Another (Boyd and Johnson, 1968), is that the androgen given during the critical period increases the sensitivity of the hypothalamo-hypophysial axis to androgen circulating later in life. Appropriately, sexual differentiation of the brain can be altered to the male pattern by the implantation of minute amounts of testosterone into the brain. Further, portions of the arcuate-ventromedial region of the hypothalamus, as well as the whole of the medial preoptic area, in androgen-treated female rats have been found to be unresponsive to electrical stimulation causing ovulation in normal females. On the other hand, exposure to light is necessary for the persistence of oestrus, for androgen-treated females kept in the dark showed fewer vaginal oestrous smears than those given access to light (Hoffman, Kordon, and Benoit, 1968). In the same study it was found that light deprivation caused atrophy of the testes, and that much greater atrophy occurred in rats given 1 mg of testosterone propionate at 5 days of age, so that it was argued that gonadotrophin secretion is more dependent upon light after testosterone-sterilization than normally. Normal animals may be able to compensate for a reduction in light, androgen-treated animals cannot. Light is also necessary for spermatogenesis in the hamster, for testicular involution occurred when only one hour of illumination daily was provided, while later comparison of the effects of different light periods showed that at least 12·5 hours of light per day were required to maintain spermatogenesis and prevent testicular degeneration (Gaston and Menaker, 1967).

The abolition of oestrous cycles, or the production of sterility, in rats can be achieved by steroids other than androgens, though in high dosage. Adrenal corticoids, progesterone, and oestrogens are all active in this way. The tranquillizing drugs, reserpine and chlorpromazine, given together with testosterone have been said to protect against the masculinizing effect of the androgen (as will the anti-androgen cyproterone acetate) and these results could be taken as evidence in favour of an action of the steroid on the brain. A single injection of oestrogen given neonatally will induce persistent oestrus, but when the administration of oestrogen is continued for up to 30 days, persistent vaginal dioestrus associated with atrophic ovaries occurs. Barraclough (1968) has suggested that the oestrogen treatment that produces persistent oestrus, permanently elevates the thresholds of responsiveness in the preoptic area so that the cyclic regulation of the ovulatory discharge of gonadotrophin is abolished and the anovulatory, persistent oestrus syndrome results. On the other hand, increase in both dosage and length of treatment may damage the tonic controlling mechanisms which may well lie in the arcuate-ventromedial area. With loss of this control, LH secretion ceases and permanent dioestrus ensues. Attention should not be directed solely to the action of steroids on the brain in arresting cycles and causing sterility after neonatal administration, despite the high importance of this response, for it has been found that the subsequent response of a number of organs to gonadal hormones is altered. For example, the oestradiol binding capacity of the uterus, as well as that of the anterior and middle hypothalamus, of adult androgen-sterilized rats is depressed below that of controls (Flerkó, Mess, and Illei-Donhoffer, 1969). Permanent changes in the metabolism of steroids by the liver also follow, as Denef and De Moor (1968) have shown in the rat. A single injection of testosterone propionate given to spayed females on the first day of life induced a pattern of steroid metabolism characteristic of males which lasted throughout life; similar treatment of castrated males prevented the differentiation of the female type of metabolism which would otherwise have occurred.

14

The Pineal Gland

The pineal gland, or epiphysis, provides a fine example of the way in which interest in a particular physiological activity can wax and wane, for until quite recently little attention was paid to the possible functions of this organ in mammals, largely because the considerable efforts of the pioneers in endocrinology had produced as much evidence against the production of a hormone as in favour. A monograph published in 1954 by Kitay and Altschule summarized the information available at that time, and this allowed the conclusion that the gland might be concerned in delaying gonadal development during infancy, for pinealectomy favoured ovarian hypertrophy and the administration of pineal extract had the opposite effect. But still the results were controversial. However, within a very few years the indole melatonin was identified in extracts of cattle pineal glands by Lerner, Case, Takahashi, Lee, and Mori (1958) and the unique location of this compound in the pineal organ has since promoted much experimental investigation. Melatonin causes the aggregation of melanin granules in the melanocytes of amphibian skin in minute concentration and may be effective at a concentration of a trillionth of a gram per millilitre of medium (Kappers, 1967). Nevertheless, the activity of this substance in mammals is much less well defined, although it may be involved in the reflex changes in gonadal function which follow alterations in environmental lighting.

Assessments of the biological function of the pineal gland, such as those of Kitay and Altschule (1954), Kitay (1967), Jouan (1968), and Wurtman, Axelrod, and Kelly (1968) have commonly dealt with its relationship to gonadal development and function, to the adrenal and thyroid glands, to pigmentation, and to a variety of other effects which include changes in brain activity, glucose metabolism, and the intestine of the foetus. Although there may be a link between pineal and pituitary gonadotrophic activity, this is not definitely established although the bulk of the positive information available, as opposed to those experiments that show no connection, indicate that the pineal gland inhibits gonadal function. However, while some attribute this action to melatonin, others consider that a peptide constituent of pineal tissue is responsible.

The pineal gland develops as an evagination from the roof of the diencephalon between the developing habenular commissures anteriorly and the posterior commissure and subcommissural organ caudally (Wurtman, Axelrod, and Kelly, 1968). In the rat it lies somewhat superficially between the cerebral and cerebellar hemispheres and beneath the confluence of the venous sinuses on the dorsal surface of the brain (Figure 14.1). In this position it can be removed without great difficulty, but in other species it is much less accessible and cannot be approached without damage to neighbouring structures. The human epiphysis lies deep within the skull and is covered by the splenium of the corpus callosum (Figure 14.1).

The pineal bodies of fishes, amphibians, and lacertilian reptiles are primarily photoreceptor organs. Degeneration may occur during ontogenesis but the ultrastructure of the pineal organs resembles that of retinal photoreceptors in many ways (Wurtman, Axelrod, and Kelly, 1968). The neurosensory cells often present in the epiphyses of submammalian vertebrates are lacking in the organs of mammalia, where the typical parenchymal element is the pinealocyte, which possesses a secretory appearance (Kappers, 1965). The type of innervation also has changed from being mainly afferent, with axons derived from nerve cells located in the pineal body, to efferent, in which the associated nerve fibres are derived from neurones situated elsewhere. Although efferent fibres enter the mammalian pineal gland from the habenular and posterior commissures, these are generally considered to be aberrant commissural loops of no functional significance, with the main supply being provided by the autonomic nervous system. Most is known about the arrangement in the rat, where two bundles of fibres, the nervi conarii, enter the dorsocaudal tip of the gland and other autonomic fibres reach the gland in company with the pial blood vessels. Within the gland the fibres in the nervi conarii break up into a diffuse network. The nervi conarii appear to be derived from the superior sympathetic chain because they degenerate, as does most of the network within the gland, after removal of both superior cervical ganglia.

Alongside melatonin, the epiphysis contains several other pharmacologically active substances, such as serotonin or 5-hydroxytryptamine, noradrenaline, and histamine. Serotonin is a precursor of melatonin (Figure 14.2), and an enzyme essential for the transformation of N-acetylserotonin to melatonin (hydroxyindole-o-methyl-transferase, HIOMT) is present only in the pineal gland. The amount of melatonin in the rat pineal gland is of the order of 0·4 µg/g and is dwarfed by that of serotonin, which is 10–20 µg/g by day

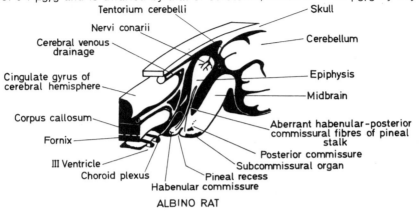

Tentorium cerebelli
Skull
Nervi conarii
Cerebral venous drainage
Cerebellum
Cingulate gyrus of cerebral hemisphere
Epiphysis
Midbrain
Corpus callosum
Aberrant habenular-posterior commissural fibres of pineal stalk
Fornix
Posterior commissure
III Ventricle
Subcommissural organ
Choroid plexus
Pineal recess
Habenular commissure

ALBINO RAT

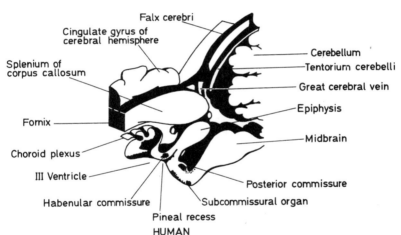

Falx cerebri
Cingulate gyrus of cerebral hemisphere
Splenium of corpus callosum
Cerebellum
Tentorium cerebelli
Great cerebral vein
Epiphysis
Fornix
Midbrain
Choroid plexus
III Ventricle
Posterior commissure
Habenular commissure
Subcommissural organ
Pineal recess

HUMAN

FIGURE 14.1 Diagrammatic representations of longitudinal slices from the diencephalic roof region of the brains of the rat and man. The facing surface of each slice has been taken at or near the median sagittal plane of the brain. From Wurtman, Axelrod, and Kelly (1968).

and 80–100 µg/g by night. Like that of the monkey, the concentration of serotonin in the gland of the rat is many times higher than the levels found in the other species so far studied (Wurtman, Axelrod, and Kelly, 1968). The concentrations of both serotonin and melatonin vary during the course of the day in several species but the peaks occur at different times. Serotonin content is lowest at night, with a peak at midday, while melatonin is low during the day and reaches a peak at the onset of darkness. The concentration of these two substances in the pineal gland also varies in phase with the oestrous cycle of the rat. A diurnal cycle in serotonin content persists after blinding in the rat, but that of HIOMT activity is abolished.

Changes in environmental lighting affect the pineal gland in a variety of ways. Exposure of rats to constant light has been shown to result in a decrease in the weight of the organ, and a change in its cytology; this response is not modified by gonadectomy, adrenalectomy, hypophysectomy, or treatment with thiouracil. It thus appears that the effect of light is relatively specific, and that the pineal changes are not secondary to the state of continuous

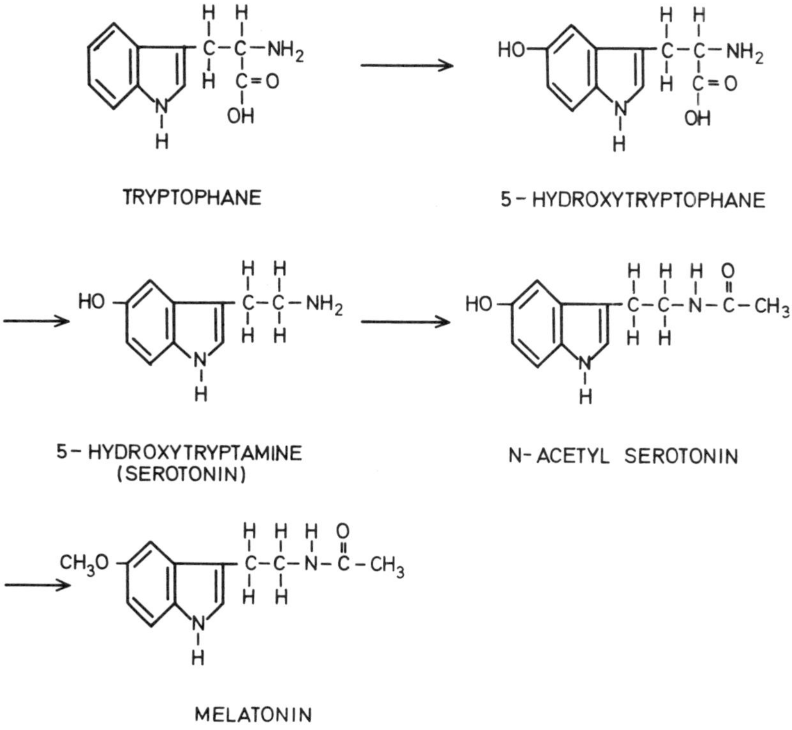

TRYPTOPHANE

5 - HYDROXYTRYPTOPHANE

5 - HYDROXYTRYPTAMINE
(SEROTONIN)

N- ACETYL SEROTONIN

MELATONIN

FIGURE 14.2 The biosynthesis of melatonin in the pineal gland.

vaginal oestrus and follicular activity that also sets in, but represent a primary reaction on the part of the epiphysis. This view tends to be substantiated by the fact that the HIOMT activity in the pineal gland is inhibited in rats by exposure to constant light, rises in animals kept in the dark and that the response to light is blocked by blinding or superior cervical ganglionectomy. Further, extracts of the pineal gland have been reported to inhibit the ovarian hypertrophy and prolonged oestrus induced by constant light. The pineal content of serotonin is depressed after removal of the superior cervical sympathetic ganglia or after interruption of the nervi conarii, but in this case exposure to constant light has been reported to act like cervical sympathectomy, which would not be expected if the response to light was mediated by the sympathetic nerves. Prolonged illumination also increases the activity of the enzyme making serotonin. The epiphysis is not essential for the genesis of biological rhythms, for pinealectomized rats continue to exhibit a circadian rhythm of locomotor activity when kept in constant dim light to eliminate the day-night pattern of illumination. Interestingly, removal of the pineal gland in the house sparrow completely abolishes the very regular circadian rhythm normally seen in complete darkness, but does not interfere with the entrainment to photoperiodic stimuli or light cycles (Gaston and Menaker, 1968).

The innervation of the pineal gland has assumed significance recently in view of the changes in pineal metabolism brought about by photic stimuli, and the changes in HIOMT activity upon exposure to continuous light or darkness have been utilized in work aimed at tracing the neural pathway involved in the response (Wurtman, Axelrod, and Kelly, 1968). Blinding by removal of the eyes prevented the fall in HIOMT activity normally found on exposure to continuous light but interruption of the major visual projections in the optic tracts was ineffective. On the other hand, lesions in the medial forebrain bundle did interfere with the response to continuous illumination and are believed to act by severing the inferior accessory optic tract which runs from the optic chiasma to the rostral midbrain tegmentum (Moore, Heller, Wurtman, and Axelrod, 1967). From the medial terminal nucleus of the tract (the nucleus of Bochenek) it is argued that a projection runs down the spinal cord to supply the outflow in the thoracic spinal cord to the sympathetic chain and in turn the superior cervical ganglia. Finally, the superior cervical sympathetic ganglia innervate the pineal gland (Figure 14.3). Although the HIOMT rhythm and the incidence of oestrous, prooestrous, and metoestrous vaginal smears may be disturbed by damage to the inferior accessory optic tracts or by the removal of the sympathetic ganglia, it remains to be proven that these results are not coincidental. There is a diurnal rhythm in the noradrenaline content of the rat pineal gland, with the concentration being highest at the end of the night and falling steadily during the day. The rhythm is abolished by blinding, by exposure to continuous darkness or light, or by dividing the preganglionic fibres to the superior cervical ganglia, just as is the oscillation in HIOMT activity. The parallel between HIOMT activity and noradrenaline content of the pineal gland can be carried further, in that interruption of the medial forebrain bundle, and fibres of the inferior accessory optic tract, arrests the fluctuations in both (Wurtman, Axelrod, and Kelly, 1968). However, such changes may appear less significant when it is realized that the noradrenaline content of the salivary glands also varies rhythmically in phase with that of the epiphysis, and that the patterns observed may simply reflect an overall diurnal periodicity in the activity of the sympathetic system as a whole. It is also not clear how the pineal gland can influence pituitary function. Does melatonin, or another pineal factor, diffuse from the gland to the hypothalamus, or is melatonin released into the general circulation to act upon the brain, for radioactive melatonin is concentrated in the hypothalamus and midbrain. Since melatonin can act as an anti-metabolite to serotonin it is conceivable that the pineal gland could influence the serotoninergic input to the hypothalamus.

Pineal function has been most closely linked with that of the gonads because pineal gland tumours are frequently found in clinical cases of precocious puberty. Nevertheless, a link between the pineal gland and the timing of puberty in children has never been established although it may be of significance that tumours which are associated with pubertas praecox are mostly non-parenchymatous growths that destroy the gland, whereas parenchymatous pinealomas less commonly have this effect. However, involvement of the pineal gland is not always found in cases where brain tumours are associated with precocious puberty. Experimental evidence for a relationship between

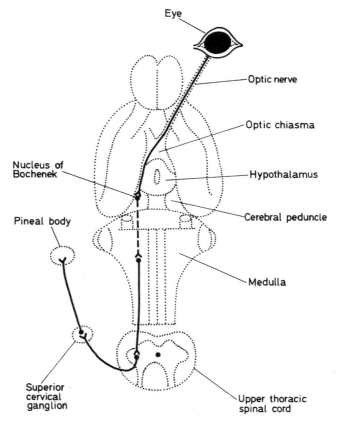

FIGURE 14.3 Schematic diagram of the pathway possibly taken by light impulses between the retina and the pineal gland of the rat. No information is available about the locus of participating fibres between the nucleus of Bochenek (the medial terminal nucleus of the inferior accessory optic tract) and the preganglionic fibres to the superior cervical ganglia. From Wurtman, Axelrod, and Kelly (1968).

pineal gland activity and that of the testes or ovaries is based largely on the responses to pinealectomy which include gonadal hypertrophy, acceleration of vaginal opening in immature rats, and the prolongation of periods of vaginal oestrus. Conversely, pineal extracts most often have been found to depress gonadal and pituitary gonadotrophic activity (Wurtman, Axelrod, and Kelly, 1968). It has been suggested (Motta, Fraschini, and Martini, 1967; Martini, Fraschini, and Motta, 1968) that while the pineal gland inhibits the secretion of both FSH and LH, melatonin acts only to suppress the secretion of LH, for removal of the pineal gland of male rats caused a significant increase in the weight of the testes, prostate gland, and seminal vesicles which is indicative of an enhanced secretion of both FSH and LH. On the other hand, melatonin given to male rats reduced prostate and seminal vesicle weight but did not cause testicular atrophy, so that it was presumed that the indole suppressed the secretion of LH without disturbing FSH release. Implants of melatonin made into the median eminence of castrated male rats reduced the pituitary store of LH and depressed the plasma level of the hormone. Melatonin was also found to delay puberty in the female rat but in this case ovarian weight was reduced, as well as that of the uterus. Here it is hard to eliminate an affect on FSH secretion. Perhaps the most convincing evidence for the participation of the pineal gland in the control of gonadal function comes from work on the hamster, where exposure of males to but one hour of light daily, normally causes atrophy of the testes and accessory sexual organs. However, after removal of the pineal gland, limitation of illumination to one hour each day fails to depress gonadotrophin secretion (Hoffman and Reiter, 1965). Blinding has a similar effect to that of habitual darkness in causing gonadal atrophy, but not when concurrent pinealectomy or superior cervical ganglionectomy is performed. Performance of either operation after 9 weeks of darkness allowed the involuted testes and accessory organs to regain their normal weights over a period of about 8 weeks (Reiter, 1968). In the female hamster, ovarian weight is not depressed by blinding but the uterus is reduced in size; the decline in uterine weight is blocked by ablation of the epiphysis. The studies on the hamster indicate that the pineal gland may be involved in the gonadal atrophy that follows deprivation of light, and pineal-ectomy has also prevented the acceleration of oestrus in ferrets which usually occurs upon exposure to extended illumination in winter, but does not interfere with the onset of oestrus in the following spring (Herbert, 1969).

In 1966, current perspectives and concepts in endocrinology were dis-cussed by Greep, who considered that 'the pineal body, after knocking at the gates for many years, has finally been admitted to full membership of the endocrine family'. He went on to remark that 'meaningful explorations of the possible role of the pineal as a modulator of the influence of such cyclic phenomena as day length, seasons and circadian rhythms on endocrine activity have only begun. The field holds promise of new developments and the further identification of functional interdigitation of the endocrine and autonomic nervous system'. It remains to be seen whether these predictions prove accurate or overoptimistic.

15

Control of Hunger and Thirst

Although the control of food and water intake does not customarily fall within the compass of neuroendocrinology, discussion of the mechanisms involved is helpful in illustrating how neurological and hormonal processes become integrated, to highlight the marked overlap of function within parts of the hypothalamus, and to illustrate some physiological or neuroendocrinological interactions.

The Regulation of Water Intake

Both the ingestion and elimination of water are controlled by the hypothalamus. Attention was drawn to the possible existence of osmoreceptors within the hypothalamus by the work of Verney (1947) on the release of antidiuretic hormone after the injection of hypertonic saline into the carotid artery of the dog. Verney was particularly concerned with the control of kidney function, but other workers, who were more interested in the mechanisms controlling thirst, soon realized that these, or similar receptors, might also be involved in the control of fluid intake. It has since emerged that the hypothalamic processes involved in thirst are separable from, though integrated with, those governing the release of antidiuretic hormone.

Wolf (1950) first postulated that osmoreceptors for thirst might exist, because the threshold of thirst to slow infusions of hypertonic sodium chloride was similar to that for the release of antidiuretic hormone. Participation of hypothalamic osmoreceptors in the control of drinking became very clear in the course of a classical series of studies on the goat made by Andersson (1953, 1966), who showed that when 0·1 ml of 1·5–2% sodium chloride solution was injected into the hypothalamus of conscious goats, drinking was initiated. The injection of hypotonic or isotonic salt solution, by contrast, did not cause drinking, and the effective area was near the descending columns of the fornix and the paraventricular nuclei. Despite the unequivocal results obtained in this work it proved difficult to carry out reproducible studies on an individual goat, or to obtain a very precise localization of the sensitive area because of the rapid diffusion of the salt solution away from the injection site. This complication was overcome by electrical stimulation of the hypothalamus, for excitation of the antero-medial hypothalamus by Andersson and McCann (1955) duplicated the response to the injection of hypertonic saline and elicited a much more prolonged bout of drinking. Ingestion of water would continue as long as excitation was maintained and could reach as much as 40% of the body weight of the goat, with consequent haemodilution and some haemolysis. Drinking could be obtained by stimulation of a lateral region lying between the descending columns of the fornix and the mamillo-thalamic tracts, and which was somewhat caudal to that reacting to the injection of hypertonic saline. It is particularly noteworthy that stimulation of the paraventricular nucleus caused antidiuresis and milk ejection, but not drinking. Stimulation of the hypothalamus of the pigeon, dog, and rat, has also induced drinking, but in the rat the sensitive area lies lateral to the fornix near the dorsomedial nucleus. Work on the effects of self-stimulation of the hypothalamus of the rat has shown that subjects with suitably placed electrodes like to promote drinking. One animal excited its lateral hypothalamus continuously for 24 hours and drank 460 ml water. Occasionally it stopped pressing the lever connected to the stimulator while pausing to eat (Mogensen and Stevenson, 1967). Neurochemical stimulation of the hypothalamus can also cause drinking and Grossman (1964) found that the repeated application of 1–5 µg of solid drug elicited reproducible effects. The placement of 1–3 µg of acetylcholine chloride into the lateral hypothalamus of satiated rats induced vigorous and prolonged drinking which began 4–8 minutes after central stimulation. The duration of the effect was extended by mixing a cholinesterase inhibitor with the acetylcholine, or by using carbamylcholine chloride, which is not hydrolyzed by cholinesterase. With carbamylcholine, vigorous drinking persisted for 30–40 minutes. In the early experiments, the comparable application of adrenergic substances to the lateral hypothalamus caused eating, but not drinking, and the specificity of the response to cholinergic or adrenergic substances seemed to be confirmed by the results of experiments in which the appropriate pharmacological blocking agents were also given. However, drinking has occasionally been induced by the injection of noradrenaline into the lateral hypothalamus of the rat, and cholinergic agents do not appear to be effective when applied to the hypothalamus of the cat or monkey. In fact, cholino-mimetic compounds applied to the lateral hypothalamus blocked eating and drinking in a rhesus monkey that had been deprived of food and water for nearly a day. The effect was dose-dependent and could be completely reversed by the microinjection of atropine at the same time. Adrenergic stimulation could elicit dose-dependent eating which was usually accompanied by drinking (Myers and Sharpe, 1968). Interestingly, the placement of small amounts of sodium chloride into the lateral hypothalamus (osmotic stimulation) did not significantly affect water intake in the rat.

Lesions in the lateral hypothalamus may cause adipsia, and aphagia, in the rat. If left untreated, such experimental rats will thirst and fast unto death, but if food and water is administered by stomach tube for some time post-operatively, the voluntary ingestion of food is resumed. Water is also drunk, but only in the course of a meal of dry food. This prandial drinking does not occur in the normal rat and is believed to be stimulated by the passage of dry food through a dry mouth, for a similar drinking pattern has been observed in rats deprived of saliva (Kissileff, 1969). Osmotic balance is of no importance in this respect, for prandial drinking is not enhanced by water deprivation, the injection of hypertonic saline, or heating. Adipsic dogs have been studied which refused water but readily accepted diluted milk or water flavoured with meat extract, when it seemed that the liquid was regarded as food.

Lesions in the preoptic area cause permanent loss of the urge to drink, as has been well shown in the goat (Andersson, 1966). Such lesioned animals retained the urge to eat and would consume liquid food, but were quite un-interested in water despite being severely dehydrated. Cooling of the preoptic area has also been found to reduce water intake in the goat and, conversely, local warming of the preoptic-anterior hypothalamic region induced drinking, though sometimes after a long latency and with drinking continued after heating was stopped.

Neural structures concerned with drinking are not restricted to the hypo-thalamus. Interference with the amygdaloid region, for example, can bring about a sharp rise in water intake, but the various components of the amygdala act differently. Thus water intake may increase with damage to the postero-ventral part of the amygdala, but falls when the lesions are placed more anteriorly, so that it could be that the anterior region of the ventral amygdala facilitates drinking, whereas the posterior part is inhibitory (see also p. 161). A significant aspect of the activity of the amygdala is that it serves to modulate the tone of the basic mechanism, for changes in the expected fluid intake are usually seen only in thirsting or deprived rats. The drinking of satiated rats is little affected.

Localized cholinergic stimulation of many other parts of the brain has caused drinking in rats (Fisher and Coury, 1964; Coury, 1967). Positive responses have been obtained from the dorsomedial hippocampus, septal region, preoptic area, mamillary body, anterior thalamus, and cingulate cortex, and these correlate very closely with the component parts of the Papez and Nauta circuits (p. 25). Cholinergic stimulation of units of the limbic system by the application of carbachol to the antero-dorsal hippocampus, cingulum, or fornix, has caused prolonged drinking in rats. The occurrence of functional localization within the hippocampus is indicated by the observation that stimulation of the postero-ventral hippocampus had little effect, and that the latencies of minutes observed before drinking began after cholinergic stimulation of different sites within the hippocampus, may have been due to diffusion of the drug along the guide cannulas (Grant and Jarrard, 1968).

Large lesions in the midbrain reduce water intake in cats and dogs. This effect has been attributed to destruction of the subcommissural organ, which is located in the roof of the cerebral aqueduct just ventral to the posterior commissure and the pineal gland. Extracts made of the organs derived from rats, cattle, or pigs, reduced water intake in rats and dogs (Gilbert, 1964), but the conclusion that the subcommissural organ is involved in the regulation of water intake has proved controversial (Crow, 1964). However, it is note-worthy than an increase in nuclear volume in the cells of the subcommissural and subfornical organs has been observed 30 minutes after bleeding in rats, and was paralleled by a smaller change in the nuclear volume of the cells of the paraventricular nuclei. Increase in the blood volume of rats by the intragastric administration of physiological saline brought about a decrease of nuclear volume in the subcommissural and subfornical organs 30 minutes later, with lesser shrinkage being observed in the cells of the supraoptic and para-ventricular nuclei (Palkovits, Zaborszky, and Magyar, 1968). While not definitive, such findings may also indicate that the subcommissural and subfornical organs serve as volume receptors.

The information stemming from analysis of the role of the diencephalon in the control of thirst is clearly incompatible with the view, championed by Cannon, that the sensation arises in the mouth or throat, and is related to the degree of moistness of the mucosa (Grossman, 1967a), although the occur-rence of prandial drinking in the brain-damaged rat shows that in special circumstances oral sensations can be important. On the other hand, the possible significance of dehydration and an increase in the osmotic pressure of the blood has been considered for a hundred and fifty years or more: the difficulty, until recently, lay in the provision of experimental proof that changes in the tonicity of the fluid perfusing the brain could be influential. With the deprivation of water, a rise in the tonicity of both intracellular and extracellular fluid ensues, as well as a reduction in the volume of both compartments. Thirst follows the administration of sodium salts and seems to be due to a decreased intracellular fluid volume, for there is no absolute dehydration but rather a movement of water from the intracellular to the extracellular fluid compartment (Andersson, 1966). The urge to drink may be traced to the same processes controlling the secretion of ADH, for there is evidence that thirst sets in with a 1–3% fall in cellular water content, which closely compares with the 1–2% depression needed to trigger ADH release. Nevertheless, it is necessary to recall that thirst arises when the volume of body fluids is reduced without changing their osmolarity, such as after haemorrhage. Release of antidiuretic hormone also occurs after loss of blood and it is possible that the osmoreceptors located in the region of the liver (p. 58) may also serve to promote drinking.

The most recent advances in understanding the factors concerned with the control of drinking implicate the kidney and the renin-angiotensin system (p. 78), for drinking after haemorrhage is considerably reduced after nephrectomy. Fitzsimons (1969) has shown that constriction of the aorta above the renal arteries in rats caused drinking, whereas constriction of the aorta below the renal arteries was ineffective. Ligation of the vena cava also promoted drinking, but less water was drunk after removal of the kidneys.

It appeared from such experiments that changes in the renal circulation alter the intake of water, and that the kidney produces a thirst factor, or dipsogen. Extracts of kidneys are known to increase the intake of water in nephrectomized animals, and in the work of Fitzsimons (1969) it proved impossible to separate the dipsogenic from the pressor activity of kidney cortex, so that it seemed likely that renin was the active constituent. In accordance with this conclusion, the infusion of the renin derivative, angiotensin, into rats rapidly promoted drinking, as did the direct injection of minute amounts of angiotensin into the anterior hypothalamic and medial preoptic area of the brain. If, as is likely, this concept is substantiated, the processes involved in the control of water intake and excretion provide a fine example of endocrine co-ordination (Figure 15.1), for not only is the operation of several endocrine mechanisms brought into phase, but the system can be extended to involve the reabsorption of sodium by the adrenal cortex and consequent water retention.

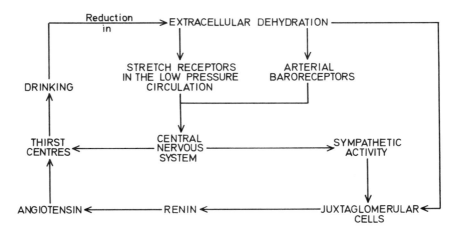

FIGURE 15.1 Some physiological interactions evident in the mechanisms concerned with the control of drinking. After Fitzsimons (1969).

Regulation of Food Intake

The fact that the body weight of an individual tended to remain constant despite fluctuations in daily food intake attracted the attention of Seguin and Lavoisier in 1790, who in discussing energy metabolism remarked: 'One can allow himself to admire the system of general liberty that nature seems to have established in everything related to living beings. In giving them life, spontaneous movement, active force, specific needs and emotions, she has also allowed for use of the same. She has even wished them to be free to abuse the same, but wisely has provided regulators and has interposed satiety after enjoyment.' (Adolph, 1961). As will emerge, several of these important regulators are in the hypothalamus.

On the premise that food intake was controlled through the sensation of hunger, which in turn involves the concept of satiation, gastric activity was first believed to be of prime importance (Anand, 1961; Mayer and Thomas, 1967). Essentially, it was suggested that eating was prompted by stomach contractions and that satiation was marked by cessation of stomach movements. However, this concept did not withstand experimental test, for removal or denervation of the stomach did not stop eating or cause loss of the sensation of hunger, and an insufficiency of nutrients after dilution of food with inert material was quickly corrected for by increasing the intake. Further, postulations that a hunger centre existed in the brain that was sensitive to the concentration of nutrient materials in the blood, were attracting attention at the turn of the century, and received support from the association of adiposity with disease processes involving the diencephalon. Twenty-five years later, improvements in the techniques of hypophysectomy and the experimental placement of lesions within the brain showed that adiposity could be produced in hypophysectomized rats by hypothalamic lesions, and it has since been shown that the disorder lies in a greatly increased food intake or hyperphagia which greatly exceeds bodily requirements (Brobeck, 1946). The limitation of the food consumption of lesioned animals to that of normal control rats by paired feeding techniques (in which the experimental animal is supplied only with the amount of food consumed by its control on the previous day), prevented obesity.

Much work has been done on hyperphagic animals and there is general agreement that the effective lesions are located in, or ventrolateral to, the ventromedial nuclei of the hypothalamus. Interference with the ventromedial nuclei in other ways has been shown to alter food intake in that the production of chemical lesions by injecting gold thiolglucose into mice causes obesity, and the bilateral anaesthetization of the zone caused the resumption of feeding in satiated rats and hyperphagia in ruminants (Mayer and Thomas, 1967). Since lesions in the fornix, hippocampus, and preoptic area have also been observed in mice treated with gold thiolglucose, it has been argued that the drug is not specific in action but simply attacks those areas accessible because of an increased permeability of the blood-brain barrier. This possibility was examined by Mayer and Arees (1968) who gave gold thiolglucose to mice after deliberately damaging cerebral cortical tissue to enhance its permeability. No additional harm was caused by the drug. Caution has also been advised in relating hyperphagia solely to destruction of the ventromedial nuclei because radio frequency lesions induced obesity less frequently than lesions produced electrolytically, so that irritation of the neighbouring lateral hypothalamus could be of significance (Rabin and Smith, 1968). However, lesions adjacent to the ventromedial nuclei, which should also cause irritation, do not readily produce hyperphagia, and removal of the ventromedial nuclei by suction (which would avoid long-lasting irritation) was still followed by characteristic hyperphagia.

Bilateral lesions at the extreme lateral border of the hypothalamus, in the same rostro-caudal plane as the ventromedial nuclei, produce complete adipsia and aphagia which lasts until death unless the animal is fed by stomach tube (Anand and Brobeck, 1951). This effect is independent of, and dominant to, the ventromedial nuclear mechanism, for hyperphagic animals become aphagic immediately upon sustaining damage to the lateral areas. Stimulation of the lateral areas in cats and rats increases food intake, while excitation of the medial hypothalamus has the opposite effect, though to a lesser degree. Such findings are in accord with the suggestion that the lateral hypothalamic area be called a 'feeding centre' and the medial one a 'satiety centre' (Anand, Dua, and Shoenberg, 1955). This concept should not be pressed too far, for animals with lesions in the lateral hypothalamus may resume eating if they are supplied with water and food by stomach tube to keep them alive after the operation. Recovery from aphagia is further facilitated by the provision of highly attractive solid food. Unlike aphagia, the hyperphagia ensuing after damage to the ventromedial area of the hypothalamus is permanent, although with increasing obesity it may decline to little above the normal level of food intake. In such a case, hyperphagia returns after a period when the food supply is restricted to bring about loss of body weight, and, alternatively, anorexia occurs in hyperphagic rats after a period of forced feeding, until much of the additional, involuntary, weight gained is lost. Study of the behaviour of hyperphagic rats has shown that the deficit lies not in the rate of feeding of the animals, which is not greatly altered, but in the period of time spent eating, which is greatly extended. Put another way, the rats do not readily become satiated, and so eat bigger meals. Hyperphagic animals are not especially highly motivated toward food and do less well than normal animals in tests designed to measure the urge to eat. When obese, hyperphagic rats become highly discriminating in their choice of food so that consumption of the routine diet falls. But as soon as delectable meals are provided, intake rises markedly (Mayer and Thomas, 1967). Interestingly, corresponding observations have been made in work with human obese subjects (Stunkard, 1968).

Excitation of the hypothalamus by chemical means has revealed that feeding can be elicited in satiated rats by the microinjection of noradrenaline into a number of diencephalic sites which extend from the ventrolateral to the dorsomedial hypothalamus and subthalamic region. The microinjection of adrenergic blocking agents into the same sites partially inhibited feeding in hungry animals (Grossman, 1968). In this connection, it may be noted that the anorexic action of amphetamine may result from depression of the excitability of the lateral hypothalamus.

The electrical activity of the ventromedial nuclei and the lateral areas of the hypothalamus is closely interrelated, as has been well shown in studies of the electrical activity of single units in this region of the cat hypothalamus (Oomura, Ooyama, Yamamoto, and Naka, 1967). Stimulation of the ventromedial nuclei could not suppress the single unit discharges in the lateral hypothalamus, and vice versa, but it was frequently observed that when the rate of discharge of cells in the ventromedial nucleus was high, that in the lateral hypothalamus was low. Additionally, stimulation of one ventromedial nucleus facilitated the electrical activity of its partner, while stimulation of the lateral hypothalamus on one side inhibited the activity of the opposite ventromedial nucleus. The EEG of the lateral hypothalamus changed from high amplitude slow waves, to low amplitude fast waves upon exposure to the odour of food, whereas the EEG of the ventromedial nuclei in many cases switched from low amplitude fast waves, to high amplitude slow waves, so providing unequivocal evidence of the reciprocal relationship between the two areas. Electrical stimulation of the lateral hypothalamus elicited mouth movements, chewing, and eating, which ended when the current was switched off, whereas if the ventromedial nucleus of cats was stimulated while they were eating, they stopped chewing and the food dropped from their mouths.

The anatomical study of the ventromedial nuclei by Golgi and degeneration techniques in normal animals, and in individuals with lesions of these structures, indicates that intrahypothalamic connections are abundant. Many collaterals from these nuclei make contact with neurones in the lateral hypothalamus (Szentágothai, Flerkó, Mess, and Halász, 1968). Other axons run dorsolaterally and end in the medial half of the lateral hypothalamic area. Connections to the dorsal longitudinal fasciculus or bundle of Schütz, which is the primary efferent pathway from the hypothalamus to the midbrain and vagal nuclei, are also apparent (Mayer and Arees, 1968). It is perhaps surprising that relatively large lesions all around the lateral hypothalamic 'feeding centre' have not been found to seriously affect food intake.

Changes in food intake have been observed after damage to the temporal lobes. Often, lesions in the amygdaloid area have produced hyperphagia, while occasionally, hypophagia ensued (Anand, 1961; Grossman, 1968). Thus, in one study in cats and monkeys, frontal lobe lesions including, or limited to, the posterior orbital cortex, reduced food intake, while those involving only the frontal tips and sparing the posterior orbital cortex, have led to an increase. Damage restricted to the anterior cingulate gyri had no effect. Lesions within the amygdala and periamygdala regions caused an aphagia or marked hypophagia which lasted for but a few days, whereas extensive temporal lobe lesions which spared the amygdala complex produced hyperphagia. With the additional involvement of the amygdala, an interval of a few days of marked hypophagia occurred, after which hyperphagia set in (Anand, 1961). In reviewing this field of research, Anand pointed out that the changes in food intake after damage to the limbic system were more severe in monkeys than in cats, and that the changes in cats were more marked than in rats. Generally, it also appears that interference with the limbic system disrupts eating less seriously than lesions in the hypothalamus and on this basis it was suggested that limbic structures in the frontal and temporal lobes modified food intake through an 'appetitive' discriminating mechanism, whereas the primitive urges of hunger and satiety were produced by the hypothalamus. Since that time it has become apparent that lesions in the globus pallidus and midbrain tegmentum produce inhibitory effects on feeding and drinking behaviour that may be as severe and permanent as those observed after damage to the hypothalamic regions. Lesions in the amygdala, hippocampus, frontal cortex, and thalamus, can elicit an hyperphagia as pronounced and as long-lasting as occurs after interference with the 'satiety centre' in the basal diencephalon. Additionally, both electrical and chemical stimulation of the preoptic area, septum, cingulate gyrus, and dorsal hippocampus could initiate food or water consumption (Grossman, 1967b). Further study has brought to light a complex pattern of behaviour. Grossman (1968) reported that electrical stimulation of the anterior portions of the amygdala inhibited feeding, but increased drinking, in deprived animals, whereas lesions produced hyperphagia and obesity but reduced ad libitum water intake. Damage to, or excitation of, the central portions of the amygdaloid complex did not consistently alter food or water intake but both eating and drinking were inhibited upon electrical stimulation of the posterior amygdala. Lesions in this location caused a relatively mild, but permanent, hyperphagia and hyperdipsia. Microinjection of adrenergic compounds into the amygdaloid complex did not initiate eating or drinking, but facilitated the response to food of thirsty or hungry rats. The animals worked harder to obtain food. Corresponding injections of cholinergic substances facilitated drinking so that it appeared probable that all central mechanisms related to feeding behaviour might selectively respond to adrenergic drugs, whereas all aspects of the 'thirst circuit' might respond electively to cholinergic and cholinolytic drugs (Grossman, 1968). This suggestion may be substantiated by the observation that feeding behaviour in the rat could be elicited or modified by the application of noradrenaline to most of the sites from which cholinergic drugs promoted drinking (Coury, 1967), but the validity of the observation itself is disputed (Grant and Jarrard, 1968).

The Regulating Signal for Food Intake

Since an individual normally adjusts his food intake according to his energy expenditure, and can compensate for day to day variations extremely rapidly, the neural centres governing food consumption do not function autonomously. They must react to a signal linked to the amount of food consumed and the energy expended. However, the signal utilized has proved disconcertingly difficult to identify. Since much work has been done with rodents, carnivores, and omnivores, greater consideration has been given to mechanisms possibly operative in these groups, which display many similarities. However, quite different controls may be used in ruminants, in which fermentation of the food is important and where much digestion and absorption takes place before material ingested passes into the true stomach (Baile, 1968).

The participation of some blood-borne signal in the control of food intake is indicated by work with parabiotic rats in which one partner was made hyperphagic by lesions of the hypothalamus. While the lesioned animal became obese, its partner exhibited anorexia and wasted away. In experiments involving the direct mixing of the blood of satiated and deprived rats, the food intake of the deprived animal was reduced by 50%, although that of the satiated animal was not affected. Thus there is evidence for a humoral factor in the blood that inhibits food intake, but not for one that exerts an excitatory action (Davis, Gallagher, and Ladove, 1967). Almost the ultimate refinement in such humoral experiments, although directly derived from the classical procedures of Dale and Loewi, has taken the form of perfusing the hypothalamus of one monkey with a saline solution and transfusing the perfusate immediately to the corresponding site in a recipient assay animal. Fluid collected from the lateral hypothalamus of a hungry donor and applied to the lateral hypothalamus of a satiated monkey caused the recipient to eat (Myers and Sharpe, 1968).

There is much evidence to support the view that the blood glucose level is important in the control of food intake, or of satiety (Mayer and Thomas, 1967). One important feature is the action of gold thioglucose in damaging the ventromedial nuclei of the hypothalamus and causing obesity in mice, and the finding that the administration of compounds in which the goldthiol moiety is joined to metabolites other than glucose, such as gold thiomalate, gold thiogalactose, gold thiosorbitol, and gold thioglycerol, do not cause ventromedial damage or obesity. This result strongly suggests that the cells of these nuclei have an especially great affinity for glucose. The electrical activity of the medial and lateral areas of the hypothalamus, but not of other parts, changes with alterations in the blood glucose level. Hyperglycaemia increases the activity in the ventromedial area and depresses that of the lateral field, whereas hypoglycaemia causes the converse. Such reactions do not follow the infusion of amino acids or fats. In starved animals, Anand, Chhina, Sharma, Dua, and Singh (1964) found that the unit activity of neurones in the 'satiety centre' was slower than that of neurones in the 'feeding centre'. When glucose was given intravenously the frequency of the spikes recorded from the 'satiety centre' increased, while that of the lateral zone fell. A reverse pattern of response occurred after depression of the blood glucose level by the administration of insulin, and no significant changes were observed in other hypothalamic or cortical neurones so that the reaction of the basal hypothalamic cells seemed to be specific. It proved difficult to show a good correlation between the activity of the neurones of the 'satiety centre' and the blood glucose level and a closer match was obtained between unit activity and the arterio-venous glucose difference, which is an index of the utilization of glucose by the brain. Autoradiographic studies with labelled glucose indicate selective uptake by the hypothalamus, but it remains odd that the direct application of glucose to the hypothalamus has not consistently inhibited feeding. Perhaps this can be accounted for by the extremely rapid diffusion of the carbohydrate away from the injection site. The discharge rate of neurones or units in the ventromedial or lateral hypothalamus has been shown to change upon the direct application of glucose (Oomura, Ono, Ooyama, and Wayner, 1969). An essential part of the neurological system that may promote feeding in response to a fall in blood sugar may be located in or near the lateral hypothalamus, for rats that have been nursed through the immediate aphagia and anorexia following lateral hypothalamic lesions and show self-regulation of food intake, do not increase input during insulin-induced hypoglycaemia, although they do increase food intake in response to dilution of the diet. On the other hand, hyperphagic rats with damaged ventromedial nuclei eat more in response to hypoglycaemia so that the glucose-sensitive area would seem to spread outside these nuclei (Epstein and Teitelbaum, 1967).

Despite many demonstrations of glucose-sensitive neurones in the basal hypothalamus, control of food intake cannot be solely governed by the concentration of glucose in the blood. Hyperphagia can be present in diabetics despite the high blood glucose levels, so that the availability of glucose to the nerve cells may be of greater significance. The fact that part of the hypothalamus, unlike other structures in the nervous system, requires insulin for the entry of glucose into cells, provides an explanation for the hyperphagia in cases of diabetes mellitus. The satiety signal (glucose) cannot enter the appropriate neurones. The availability or utilization of glucose has been expressed in terms of the difference in concentration between arterial and venous blood across the brain, and it appears that hunger is experienced when the arterio-venous difference tends toward zero, with satiety prevailing when the difference is appreciable. In hungry subjects, the intravenous injection of glucose raised the blood glucose concentration and the arterio-venous difference and there was a loss of the sensation of hunger. In diabetic subjects, by contrast, intravenous glucose again raised the blood concentration but not the arterio-venous difference and there was no loss of the feeling of hunger or of stomach movements. The metabolism of glucose and lipids is so interrelated that Kennedy (1966) argues that it is difficult to imagine a feedback to the hypothalamus of information about one nutrient that is not affected by the other, whether over the short- or long-term, so that the plasma level of non-esterified fatty acids could provide an important satiety signal. To some extent, a feedback action of fatty acids upon the hypothalamus could account for the common finding that the food intake of hyperphagic rats falls once a certain amount of fat has accumulated. Indeed, this phenomenon has led Mayer to suggest that the short-term regulation of food intake is glucostatic in character, whereas the long-term is lipostatic. However, Kennedy (1966) has observed that the plasma level of fatty acids can be high both in hunger and in obesity, and inclined to the view that the plasma insulin level was possibly of greater significance, for this was high after feeding, and became higher still once obesity occurred.

There is evidence that the plasma concentration of amino acids may be influential in the control of food intake, but it seems curious that when a surplus of all but one of the essential amino acids is added to a low protein diet, food intake is depressed. Some indication that receptors exist within the brain, that are sensitive to the concentration of the missing amino acid, has been provided by experiments in which the required amino acid was infused into the carotid artery or jugular vein of rats for periods measured in days. When threonine, for example, was given through the carotid artery for several days, the food intake rose, but when the jugular vein was employed, food consumption remained low (Leung and Rogers, 1969).

In spite of the mass of information concerning the role of the hypothalamus in the control of food intake, it has proved difficult to account for the fact that meals usually end long before significant absorption, and before the metabolites derived from the food could modify neural function. This phenomenon may be accounted for by the occurrence of receptors in the stomach that are sensitive to the nutritive value of ingested foods (Mayer and Thomas, 1967). The placement of glucose directly into the stomach of rats produced almost immediate satiety, and animals provided with permanently implanted gastric tubes could adjust their oral intake extremely quickly to compensate for the calorific value of liquid nutrients injected directly into the stomach. The intraperitoneal injection of glucose into rats or cats with lesions of the hypothalamus causing hyperphagia was still able to cause anorexia, despite loss of major central glucose receptors. Increased electrical activity of the 'satiety centres' with distension of the stomach, or upon the application of 5·4% glucose, or a solution of glycine, increased the firing rate of mesenteric nerves from isolated loops of the gut in dogs (Sharma, 1967).

The blood level of glucose is influential in the control of several neuroendocrine mechanisms. The pituitary secretion of growth hormone and of ACTH is affected, as is the release of catecholamines from the adrenal medulla. As yet it has not proved possible to separate the hypothalamic components responsible for one process from those of another and there must be a high degree of functional overlap, for the ventromedial nuclei are involved in all. In this respect the limitations of present ideas on the function of the ventromedial nuclei are illustrated by the difficulty caused by the observation that the degree and frequency of hyperphagia after damage to the nuclei is greater in female rats than in males (Cox, Kakolewski, and Valenstein, 1969). This sex difference can be attributed, at least in part, to a different hormonal background but this postulation raises many more questions about the mode of action of hormones on this part of the brain.

A remarkable parallel is demonstrable between recovery from the effects of damage to the lateral hypothalamus and the development of control of food intake by rats thyroidectomized at birth (Teitelbaum, 1967; Teitelbaum, Cheng, and Rozin, 1969). When tested at 21 days, which is the normal age at weaning, the hypothyroid rats displayed every symptom of the lateral hypothalamic syndrome. The retarded weanling rat was completely aphagic and adipsic when offered food and water, although it took milk from the mother. More fully developed individuals accepted wet and palatable foods but did not eat enough to promote survival. As development proceeded, voluntary feeding began and the caloric intake of a liquid diet could be controlled, but the animals remained adipsic and died if only dry food and water were available. Later, when dry food and water was accepted, drinking was prandial and could not be stimulated by dehydration. To carry the parallel with the lateral hypothalamic injured animals even further, eating was not increased in response to hypoglycaemia (see p. 163).

16

Hormones and Brain Function

The relationships between hormones and brain function are many and varied. Those related to the control of secretion of pituitary hormones have been covered in earlier chapters. Here, more general interactions between hormones and nervous activity will be discussed, although one major aspect, that of hormones and behaviour, will be largely neglected since it is too vast a field to be covered in synoptic fashion. Nevertheless, many features of the actions of hormones on neural tissue are particularly well illustrated by their behavioural effects. For example, classic work on the cat indicates that the hypothalamus inhibits the manifestation of sexual behaviour and that oestrogen relieves or removes the inhibition (Bard, 1940). Essentially, the cat in oestrus differs from that in anoestrus or after spaying in reacting to vaginal stimulation by lordosis and treading, whereas similar stimulation of the ovariectomized cat causes escape or anger. However, after transection of the spinal cord or of the midbrain, genital stimulation elicits a typical oestrual response in the spayed female and treatment with oestrogen does not alter the intensity of the reaction. Under these circumstances, sensitivity to the presence or absence of oestrogen became demonstrable only if transection of the brain was carried out above the hypothalamus so that its projections with the spinal cord remained intact. In the light of this study the central nervous system is evidently not uniformly sensitive to hormones, so that the part played by the hypothalamus becomes of special importance.

Steroids and Brain Excitability

A variety of observations establish that steroid hormones are important in setting the level of brain excitability (Sawyer, 1966; Woodbury and Vernadakis, 1966). Variations in brain excitability can be measured by determining the threshold for electrically-induced seizures (the electroshock seizure threshold, EST), by study of the change in threshold for Metrazol, strychnine or insulin-induced seizures, by study of the susceptibility of an individual to audiogenic seizures, or by following the occurrence of spontaneous seizures. Adrenocortical insufficiency, whether in patients or experimental animals, is associated with the appearance of convulsions, and seizures can be induced more readily. Sensations of nervousness, irritability, depression, and hallucinations are experienced by patients, and diffuse slow wave activity is apparent in the EEG. The sense of taste is also greatly enhanced in Addisonian patients and becomes depressed to the normal level upon treatment with a glucocorticoid. These effects do not arise primarily because of the lack of adrenal hormones, but as a result of the changes in electrolyte concentration in the plasma and brain which accompany adrenal hypofunction. Accordingly, they can be enhanced by supplying large amounts of fluid to cause water intoxication. An increased concentration of sodium is present in the cells of the brains of adrenalectomized animals so that the ratio of extracellular to intracellular sodium falls; the turnover of sodium is also said to be decreased. These changes seem to follow damage to the intracellular pump which normally lowers the sodium content of the cell, and related changes in potassium concentration ensue. The total amount of potassium within the brain is unchanged but the ratio of intracellular to extracellular potassium concentration is decreased. Along with these alterations in the distribution of electrolytes comes an increase in brain excitability, a slowing of the frequency of the EEG, and a decrease in central conduction time.

As might be expected, adrenocortical steroids cause characteristic changes in brain excitability, and the changes observed vary with the kind of steroid supplied. In large doses, some steroids, including deoxycorticosterone and progesterone, have sedative effects which can progress to anaesthesia, although the amounts required for the latter reaction are exceedingly high. Steroids that predominantly affect electrolyte metabolism (deoxycorticosterone, 11-deoxycostisol) raise the threshold for electrically-induced seizures, whereas those that are mainly concerned with carbohydrate and electrolyte metabolism (cortisone, cortisol) increase excitability. Other steroids (corticosterone, 11-dehydrocorticosterone) affect both carbohydrate and electrolyte metabolism, and these fall between the other two groups in affecting the excitability of the brain. The excitatory action of cortisone or cortisol is not mediated by changes in electrolyte metabolism, since this is not greatly altered with chronic administration of the hormones. The mechanism of action of these substances on the brain is not understood.

Removal of the gonads has a minor effect on brain excitability when compared to that of adrenalectomy, and tends to render the individual less susceptible to seizures. Treatment with oestradiol in physiological dosage greatly increases brain excitability. In female rats, excitability is higher than in males, although it varies in phase with the oestrous cycle and is highest at oestrus. Brain excitability is slightly decreased by testosterone or progesterone.

Adrenal hormones are important in sustaining the flow of blood through the brain. Hoagland (1954) studied the effect of adrenalectomy upon cerebral blood flow and upon the consumption of oxygen by brain tissue, and found that in rats there was a fall of 57% in cerebral blood flow and an 18% drop in oxygen consumption. Both parameters were restored to normal upon treatment with cortisone. At least some of the electrical changes produced by adrenalectomy may be attributed to such a reduction on brain blood flow.

After thyroidectomy, brain excitability is decreased and it is more difficult to induce seizures. This change has been related to a decrease in intracellular sodium concentration. As after adrenalectomy, the frequencies of the electroencephalographic rhythms fall. Conversely, chronic treatment with thyroxine or tri-iodothyronine lowers the electroshock seizure threshold, raises the concentration of sodium within the nerve cell, and increases the alpha rhythm of the EEG in man (Woodbury and Vernadakis, 1966).

Hormones and Brain Development

Critical Periods. The administration of steroids for a brief period of time early in life can have long-lasting effects on the secretion of pituitary hormones, and on behaviour. For the greatest modification of adult brain function the hormone must be given during a certain stage of development, which is often called a critical period. For example, the critical period for androgen sterilization of the female rat (chapter 13) occurs neonatally, and covers approximately the first ten days after birth, when growth of brain cells is taking place and their axonic and dendritic connections are proliferating. From the point of view of the brain, birth is an unimportant event. Indeed, the degree of brain growth at delivery is more closely related to the length of gestation, for the brain of the neonatal guinea-pig is much more highly developed than that of the rat. It is thus better to distinguish several phases in the differentiation of the brain and to limit comparisons between species to brains at equivalent stages. One such scheme is that of Davison and Dobbing (1968), which utilizes four stages. These are:

Stage I, which covers the embryological formation of the brain and ends when multiplication of the neuroblasts ceases. An adult complement of neurones is present and the nuclear pattern of the brain laid down.
Stage II, is the period when the size of the brain rapidly increases and is marked by the growth of axons and dendrites and the establishment of neuronal connections, as well as the multiplication of oligodendroglial

cells and the deposition of myelin sheaths by them. These events may occur early in stage II and have been treated as stage IIa, with a period of general brain growth providing stage IIb.

Stage III, accommodates the growth in size of the brain which accompanies general body growth in the young; the transition from stage II is not readily determined. The adult brain remains in this condition until senility sets in and Stage *IV*, which is the period of degeneration and regression, is reached.

Using this scheme, the brain of a guinea-pig passes through stage IIa during the last quarter of gestation, that of a pig is in this phase of development at birth, and the brains of both species are comparable with that of a rat during the second and third post-natal week (Davison and Dobbing, 1968). The divisions outlined above are broad, and more precise measures based on the appearance of various kinds of electrical activity, the differentiation of the layers of cells in the cerebral cortex, or of some other index, may be preferable. It should also be recalled that there are regional differences in the rate of development of component parts of the brain. However, the scheme of Davison and Dobbing (1968) provides a convenient starting point.

Thyroid and Brain Development. Cretinism used to be endemic in inland, iodine-deficient, communities and provides the classic example of the way in which a lack of thyroid hormone in infancy affects growth and behaviour. In many cases the changes are irreversible despite treatment with thyroid hormone, and the reasons for this are of great interest. Most experimental work has been carried out with the rat, for it is born with a very immature brain and much is known about the development and behaviour of this species. In the young hypothyroid rat the growth rate is slowed, the opening of the eyes delayed, and hair growth is sparse. Brain function is also disturbed, for thyroidectomy at birth retards, but does not prevent, the appearance of innately-organized behaviour patterns, such as the startle, righting, and placing reactions. The capacity for adaptive behaviour later in life, as indicated by the performance in mazes and other learning tasks, is severely impaired and the extent of the impairment is directly related to the age at which the animal is rendered athyroid, and the age at which replacement therapy is initiated (Campbell and Eayrs, 1965; Eayrs, 1968; Hamburgh, 1968). It is evident that the maturation of the brain is dependent upon the beneficial influence of thyroid hormone during a restricted or critical period and the effects of a deficiency have been traced in a variety of ways. After neonatal thyroidectomy there is a delay in the myelination of nerve fibres, a significant reduction in the size of the cortical neurons, and a hypoplasia of the associated neuropil. With this reduction in the number of axons and dendrites, the probability of interaction between neurones becomes lessened; the finding that the EEG of such rats remains of low amplitude and that the electrically evoked response shows an abnormally long latency and duration, may stem from this. The maturation of the mechanisms concerned with thermoregulation is also permanently disturbed if thyroid hormone is not available at birth, and the body temperature reaches and remains at 30°C instead of the normal 36°C. Withdrawal of thyroid hormone after 10 days of age does not have such serious consequences. These deleterious changes can be reversed by treatment with thyroid hormone but the longer treatment is delayed, the more profound are the lasting effects.

The cytoarchitectonic changes in the cerebral cortex have been attributed to a reduced capacity on the part of the cells to synthesize protein, and studies both *in vivo* and *in vitro* have shown a reduction in cerebral RNA concentration per cell, and a depression of cerebral protein turnover, after neonatal thyroidectomy (Timiras, Vernadakis, and Sherwood, 1968). The effects of thyroidectomy can be ameliorated by treatment with growth hormone which, when given prenatally, promotes the proliferation of cortical neurones and increases cortical cell density and the weight of the brain. It also appears that during maturation, the mitochondria of the brain neurones lose their capacity to interact with thyroxine during protein biosynthesis. This is indicated by experiments in which mitochondria, microsomes, and cell sap were prepared from mature and immature rat brain and mixed together with isotopically labelled leucine and thyroxine, in all combinations of the three cell fractions derived from both types of brain. The degree to which the incorporation of the amino acid into protein was influenced by thyroxine depended solely upon the nature of the mitochondria present. Replacement of the adult brain mitochondria with immature brain mitochondria resulted in a thyroxine stimulation of the adult brain system, and, conversely, the replacement of immature brain mitochondria with mature brain mitochondria abolished the stimulation of protein synthesis by thyroxine that normally occurred with the immature brain system (Sokoloff, 1967).

The presence of an excess of thyroid hormone during the development of an individual can be as harmful as a deficiency. With a small excess, eye opening and skeletal maturation is hastened, but although the learning ability of the immature rat may be improved at first, the initial advantage over untreated control animals is not maintained, so that at 35–45 days of age, control rats negotiated two different kinds of maze-learning devices with less errors than littermates that had received thyroxine as infants. Schapiro (1968) has suggested that neonatal thyroxine may cause a more rapid advance through a critical ontogenetic period during which a behavioural repertoire is accumulated, and so limit the period during which adaptive responses can be readily learned. When large amounts of tri-iodothyronine are given to rats immediately after birth, more striking changes occur and growth is permanently depressed. This effect is observed only in animals treated before the 14th day of life, and the rats differ markedly from controls in that there is hypoplasia of the pinna and deformities of the snout. The adaptive behaviour of such rats, measured by the Hebb-Williams closed-field test, is also inferior to that of control rats given a roughly equivalent dose of tri-iodothyronine after the 14th day. To judge from a lowered thyroid/serum ratio of administered radioactive iodine in these animals and a longer biological half life of the isotope in the thyroid gland, the output of TSH by the hypophysis is abnormally low and the animals are chronically hypothyroid, but this explanation of the devastating effect of neonatal treatment with thyroxine is not fully satisfactory (Eayrs, 1968).

Adrenal Hormones and the Development of Neuroendocrine Function. As described in chapter 7, detectable amounts of ACTH are secreted by the pituitary gland of the foetus, although adrenal function is depressed immediately after delivery and the reaction of the pituitary gland of the neonatal individual to stresses is not fully developed. Nevertheless, the reactivity of the hypothalamo-pituitary-adrenal axis of the adult is affected by stimuli experienced in infancy. Perhaps the most extreme changes in the rat follow the injection of 1 mg of cortisol acetate on the day of birth. Growth is slowed, dental abnormalities appear, brain cholesterol and exploratory or locomotor activity are decreased, and the rate of formation of dendritic spines on the pyramidal cells of the sensory cortex is slowed. However, the pituitary-adrenal response to stress is normal in these animals when adult (Schapiro, 1968). The effect observed with hormonal treatment depends upon the stage of maturation of the responsive tissue and upon the affinity of neurological structures for specific hormones. Both cortisol and oestradiol markedly increased the sensitivity of the developing brain of the infant rat to electrical stimulation, but different critical periods were evident. Oestradiol accelerated brain maturation when given between 4 and 8 days of age, whereas the critical period for cortisol lay between 8 and 16 days. Some specificity in the action of these steroids is indicated by the fact that deoxycorticosterone failed to affect the development of the nervous system. The mechanisms by which brain maturation is influenced by the steroids is not well understood but could be mediated by changes in myelogenesis and electrolyte distribution (Timiras, Vernadakis, and Sherwood, 1968).

Adrenal function can be modified by less drastic treatment. When infant rats are stimulated by handling or other means they become less emotionally reactive than non-stimulated controls and the secretion of adrenal hormones is not promoted by environmental factors to the same degree. Animals stimulated in infancy by being picked up daily, placed in another cage for 3 minutes and then returned to the nest, showed a more rapid, and greater, secretion of adrenal corticoids to a brief, but intensive, electric shock applied when they were adult, than others not handled in this way. However, the release of adrenal hormone did not last as long as non-stimulated controls and, partly on this basis, it is argued that rats handled in infancy are more adaptable to a change in environment since they can respond to a novel situation with a moderate output of steroids, and react to a physically threatening one with a near-maximum secretion, while animals not handled in infancy can only give a near-maximal response to any change in the environment (Levine and Mullins, 1966). Handling appears to accelerate the maturation of the hypothalamic-pituitary-adrenal system, in that in handled animals as young as 3 days, a significant increase in plasma corticoid level occurs after an electric shock, while no such increase is seen in non-handled rats younger than 9 days old. Handling has also been shown to advance the onset of the circadian rhythm of adrenal function (Ader, 1969). In the particular strain of rats studied, the characteristic adrenocortical rhythm was first observed in undisturbed individuals 21–25 days old, but in animals handled for 3 minutes daily, or given an electric shock, the rhythm appeared as early as 16 days of age. Sufficient corticosterone is secreted by the rat adrenal gland in infancy to alter brain function and the output is raised by handling. Labelled corticosterone is taken up preferentially by the hypothalamus of the 2-day old rat, so that the facts available are consistent with the hypothesis that the early handling effect may be mediated by adrenal corticosterone and that this modifies the developmental pattern of the hypothalamus (Zarrow, Philpott, Denenberg, and O'Connor, 1968). Others have suggested that exposure to strong environmental stimuli, such as handling, prolongs the phase of brain maturation so that the period during which brain and behavioural mechanisms can be influenced is extended (Altman, Das, and Anderson, 1968). The evidence for this suggestion is somewhat scanty and based on differences in brain weight, measurements of brain areas, and thymidine uptake in handled and undisturbed rats. The brains of the handled animals were smaller than the controls, but cell proliferation seemed to be higher.

Gonadal Hormones and Sexual Differentiation of the Brain. The action of androgen in causing sexual differentiation of the brain in a masculine manner during a limited phase of development, and the comparable effects of over-dosage with other steroids, has been discussed in relation to the secretion of gonadotrophic hormones in chapter 13. However, the fact that comparable changes in sexual behaviour also occur should not be overlooked. Thus, male rats castrated immediately after birth are functionally female in the behavioural sense when mature, display lordosis, and allow mounting by other males when injected with oestrogen and progesterone. Conversely, the treatment of newborn female rats with androgen prevents the development of feminine sexual behaviour and favours the manifestation of masculine reactions. This is not solely due to a lack of female sex hormone, because removal of the ovaries and suitable replacement therapy with oestrogen and progesterone fails to evoke feminine responses. Neonatal castration of the male also allows the persistence of the cyclic motor activity which is typical of the female and lost in the male. However, not all of the behavioural differences between genetic males and females are lost after neonatal castration, for testosterone given to mature test rats more readily suppresses the lordosis response in males than in females (Levine, 1967).

Hormones and the Electrical Activity of the Brain

Perhaps the first indications of the action of hormones on the electrical activity of the brain came from studies of the relationship between the frequency of the α-rhythm of the EEG and the metabolic rate. In the hypothyroid state the metabolic rate and α-rhythm are both depressed, and increase upon replacement therapy with thyroxine. It has also been claimed that the analysis of the EEG can assist in distinguishing between primary and secondary hypothyroidism, and be of prognostic value in deciding whether mental recovery in cases of cretinism can be expected (Campbell and Eayrs, 1965).

Abrupt and consistent changes in the EEG of rabbits have been observed following coitus which can be mimicked by the administration of progesterone (Sawyer, Kawakami, and Kanematsu, 1966). The EEG after-reaction occurs spontaneously in female rabbits after coitus or vaginal stimulation, and includes an initial sleep-like EEG record followed by a peculiar sequence of changes with characteristics of hyperarousal which comprise high amplitude theta waves in the hippocampus, limbic cortex, and related areas and a de-synchronized frontal cortical record. Despite the electrical evidence of hippocampal hyperactivity, the rabbit appears to be in a behaviourally depressed state with partially closed eyes, constricted pupils, and slowed heart rate and respiration: a condition now termed paradoxical sleep. Low frequency stimulation of various hypothalamic and limbic areas in spayed, oestrogen-treated rabbits can also elicit the EEG after-reaction and the threshold of this response alters biphasically upon the administration of progesterone. During the first few hours after the injection the female is highly oestrous, and the EEG after-reaction threshold is much reduced, whereas 24 hours later the rabbit is behaviourally anoestrous and the EEG after-reaction threshold is highly elevated. Corresponding fluctuations can be recorded in an EEG arousal response, which follows high frequency electrical stimulation of the midbrain reticular activating system. It is of interest that certain oral contraceptives of the progesterone type also affect the threshold of the EEG after-reaction in a biphasic manner, while testosterone and the nortestosterone antifertility steroids cause only a rise in threshold; these effects may be related to the blockade of ovulation. Experiments of this type, in which the effect of HCG on the EEG of spayed rabbits was examined, provided an early indication of the direct feedback action of polypeptide hormones on the brain (chapter 10). The influence of glucocorticoids and of ACTH on the extinction of conditioned avoidance behaviour (de Wied, 1966, 1969; de Wied, Bohus, and Greven, 1968) illustrates the interaction of pituitary and target organ hormones on the brain in a different way. Essentially, this test involves training rats to perform a response to the sound of a conditioned stimulus (a buzzer) associated with the unconditioned stimulus of an electric shock. When the animals have learned to avoid the shock satisfactorily the unconditioned stimulus is no longer presented and the rate at which the rat stops taking action when the bell sounds in a series of tests measured. In the course of a fortnight or less, the number of conditioned avoidance responses in normal or blank operated rats falls to zero, whereas removal of the adrenals causes a very high degree of retention. Treatment with corticosterone facilitates the extinction of the response, so that from the results of these and other experiments it seems that ACTH and the adrenal corticoids have an antagonistic action on the brain.

Detailed examination of the changes in the EEG of the rat and rabbit brought about by the injection of progesterone, vasopressin, or adrenaline, elicited evidence that the effects observed were invariably associated with a rise in blood pressure and thus non-specific (Feyer, Ramirez, Whitmoyer, and Sawyer, 1967). The action of progesterone on the brain was explored further by Komisaruk, McDonald, Whitmoyer, and Sawyer (1967), who concluded, on the basis of the EEG changes and the unit activity affected by the hormone, that this steroid has a general inhibitory effect on brain activity. It was considered that at least some of the effects of progesterone could be mediated by a general depression of the level of arousal, and the EEG after-reaction following coitus in the rabbit, mentioned earlier, provides a fine example. On the other hand, it has been found that progesterone depresses the response of hypothalamic units to cervical stimulation, but not to pinching of the skin, so that a selective depression of the response to stimuli from the genital tract seems apparent. However, Komisaruk and his colleagues suggest that the difference in reaction is simply due to the fact that pinching causes much greater arousal than cervical probing, or gentle stroking of the fur, and so is inhibited by progesterone to a much lesser extent. Attention has also been drawn to the action of progesterone on non-sexual processes by Rothchild (1967), who points out that the inhibition of ovulation by progesterone is frequently accompanied by an increase in appetite, an increase in body temperature, and depression of motor activity, while in the oestrogen-dominated phase of the sexual cycle the tone of these processes is in the opposite direction. In his view, the inhibition is provided by depression of the restraining activity of the ventromedial nuclei upon the lateral hypothalamic nuclei which, in turn, feed inhibitory impulses to mechanisms governing motor and sexual activity, and heat loss, while promoting appetite. Much needs to be done before such a simple concept becomes fully acceptable, but this hypothesis could provide a starting point for those who wish to explore the way in which endocrine and non-endocrine mechanisms are integrated within the hypothalamus.

The electrical activity of components of the limbic system is modified by changes in the endogenous hormone pattern. In the rat, for example, the localized seizure threshold in the dorsal hippocampus has been found to fluctuate during the oestrous cycle and to be low between the morning of prooestrus and the morning of oestrus. The seizure threshold of the medial amygdala changes in a similar way to the dorsal hippocampus, but that of the lateral amygdala varies in the opposite sense. These changes are abolished by ovariectomy but reappear upon treatment with oestrogen. Progesterone increases the seizure threshold of the hippocampus and decreases that of the lateral amygdala (Terasawa and Timiras, 1968b).

Studies of the electrical activity of single cells in the hypothalamus of the rat, and of their changing response to alterations in the hormone background, and to stimulation of the genitalis, have been made (Beyer and Sawyer, 1969), but little specificity in the response to genital stimuli has been observed. Most of the neurones that respond to probing of the cervix also react to other stimulus modalities such as pain, cold, or odours. The responsiveness of the neurones has been observed to change during the oestrous cycle and there is a trend toward a reduction in excitability during oestrus. Progesterone given intravenously depresses the responsiveness to genital stimuli but this effect is short-lived and disappears an hour after injection. In pseudopregnant rats, wherein the blood level of progesterone is raised, the number of hypothalamic neurones excited by cervical probing is reduced; removal of the ovaries immediately before recording is done leads to a marked increase in the percentage of neurones excited by vaginal probing. Nevertheless, recent critical studies throw doubt on the action of progesterone on the hypothalamus (Lincoln, 1969). Fewer experiments involving the administration of oestrogen have been performed, but these are indicative of a predominantly inhibitory influence upon the response of hypothalamic neurones to manipulation of the cervix. Chhina and Anand (1969) frequently observed an inhibition of unit activity in the rostral hypothalamus of adult monkeys after genital stimulation. In the immature female, excitation of hypothalamic units predominated, but upon the administration of oestrogen this was sometimes converted to an inhibition. Units in the limbic system tended to fire more frequently in response to genital stimulation, but some were inhibited. In the cat, unit activity in the ventromedial nucleus of the hypothalamus is slightly facilitated during oestrus and inhibited during anoestrus (Kawakami and Saito, 1967). Often, mechanical stimulation of the cervix for one minute during oestrus inhibits the electrical activity of this region, which is followed by a transient facilitation when the rod is withdrawn from the vagina, and then by a marked inhibition which lasts for more than 20 minutes. The administration of LH also produces a long-lasting fall in the firing rate of units after a latency of 10–20 minutes. Analysis of the changing response of hypothalamic units in a variety of situations is complicated by the difficulty encountered in deciding whether the stimuli applied, act directly upon the cells, or through another region supplying afferents to the units. This dilemma can be avoided, to a large extent, by working with islands of hypothalamic tissue in which major extrinsic inputs can be rigorously excluded (Cross and Kitay, 1967). The spontaneous unit activity is not depressed, and changes in the level according to the phase of the oestrous cycle can be detected. The highest mean firing rate occurs at prooestrus.

The mode of action of steroid hormones on the brain remains a mystery, although some clues are available. Both oestradiol and sodium phenobarbital reduce the translocation of labelled progesterone into the brain in experiments involving the infusion of the drugs into one half of the brain, with the other serving as a control. The effect is apparent in experiments in which the appropriate solutions are given for 7 minutes only (Bidder, 1968). This work may point to one means of exercising control over hormone action —that is, by regulating the access of the material to the brain. However, it may be as well to point out that the actions of barbiturate drugs are not confined to the nervous system, for they are now known to be potent inductors of liver microsomal enzymes. Treatment of individuals with barbiturates can thus indirectly enhance the hydroxylation of steroid hormones and so upset the prevailing endocrine balances (Conney, 1967). This is illustrated by experiments in which the administration of phenobarbitone to spayed female rats for 3 days before giving oestrogen, markedly inhibited the expected increase in uterine enzyme activity (Singhal, Valadares, and Schwark, 1969). Time is required for enzyme induction, so that this process cannot account for the immediate consequences of treatment with barbiturates; but in longer-term experiments the effect could be significant.

The Blood-Brain Barrier

The existence and properties of the blood-brain barrier are often mentioned in discussions of the action of hormones on the brain, although there may be little reality in the concept (Dobbing, 1961). Originally, it appeared reasonable to postulate the existence of a physical barrier between the blood and the brain because alone of all the tissues in the body the brain remained uncoloured by dyes after systemic administration. The very fact that a few small regions such as the choroid plexus, area postrema, median eminence, and pineal gland were permeable to vital dyes added support to the idea, as did the observation that the brain became stained if the dye was injected into the cerebral ventricles. However, the barrier has not been identified although it has been equated with the astrocytic glial cells surrounding most capillaries in the brain. The reason for the non-penetration of dyes into the brain has now become apparent, in that the dyes become associated with the proteins of the blood plasma and are not free to diffuse into the brain. Since the cerebro-spinal fluid does not contain sufficient protein to absorb them, dyes such as trypan blue can diffuse through and tint the brain after intraventricular injection (Dobbing, 1968).

Since the molecules of the plasma proteins are large, it is not surprising that little penetration of the brain substance occurs, but small molecules, such as glutamic acid, also appear to be excluded. The entry of glucose, on the other hand, is facilitated, as is that of alcohol and the volatile anaesthetics. While the easy access of alcohol and the volatile anaesthetics may be due to their solubility in fats, this explanation is not acceptable for glucose; in this case it seems that passive diffusion is assisted by a carrier-mediated transport mechanism. The mechanism does not convey fructose as readily as glucose so that glucose receives preferential access, although fructose is not excluded (Dobbing, 1968). With regard to glutamic acid, the situation is somewhat different in that a transport mechanism acts in reverse to pump the substance out of the brain. When the brain is perfused with blood containing radioactive glutamic acid the overall concentration of glutanic acid in the blood may change little, but when the level of radioactivity in the brain and blood is measured, a movement of labelled compound into the nervous tissue is detected which is balanced by the elimination of a corresponding amount of the substance from the brain. Misleading conclusions may also be drawn from experiments in which the movement of a substance into the whole brain is measured. When a test substance is extracted from the entire organ, regional differences, and that between grey and white matter, are obscured and the final result is merely an average figure. This is illustrated by the observation that the uptake of radioactive glutamine by ventral horn cells is fifteen times higher than the rest of the grey matter of the spinal cord; that of lysine is six times greater. Overall, however, the uptake of glutamine by the grey matter differs little from that of the cerebral cortex. As another example, it may be pointed out that the hypothalamus has a greater affinity for catecholamines than the rest of the brain (Levin and Scicli, 1969).

Damage to the brain, as caused by focal lesions, allows the entry of substances which are normally excluded. This is a consideration to be borne in mind when evaluating the results of experiments involving the manipulation of electrodes or other devices within the skull. Chemical disruption of the barrier can also follow the application of the apparently innocuous contrast media used for cerebral angiography, as well as by metabolic inhibitors, including sulfhydryl blocking agents and heavy metal compounds (Edström, 1964).

Glutamic acid, phosphorus, potassium, and sodium ferrocyanide can readily enter the brain of the young rat until about 2 weeks after birth (Himwich, 1962), so that during this time it may be argued that the blood-brain barrier (or the processes simulating one) does not exist. However, the metabolic activity of the neonatal brain of the rat is much higher than later in life and the demand for metabolites may favour the admittance of an unusual variety of materials (Dobbing, 1968). It is also possible that the plasma proteins of the young rat bind test substances much less readily than those of the adult.

17

Current Research Trends in Neuroendocrinology

It seems appropriate to close this survey of current neuroendocrine thought with a look ahead toward the future of the subject. Every discussion of this kind is coloured by the predilections of the writer, but, with this proviso, it can be instructive to determine why certain facets of the subject, so important in hindsight, were neglected on a particular occasion.

From the anatomical point of view it seems that the time has come for the electron microscope to be used as an investigative tool, rather than as a means of peering with ever increasing precision into biological structure. With the accumulation of adequate material on the ultrastructure of normal neural and endocrine tissues, the changes brought about by experimental intervention can now be followed. Accordingly, the fine terminations of nerve fibres upon the nuclei in the hypothalamus are now being traced with the aid of the electron microscope in animals, and the results compared with the experimental situation in which lesions have been placed in areas supplying afferents to the diencephalon. The electron microscope is also proving of value in the localization and identification of neurotransmitters within individual cells, although only small fragments of tissue can be examined in this way. For the mapping of systems of neurones within the brain, histochemistry, or perhaps histomorphology, can make a special contribution. There appears to be a trend toward the classification of neural connections and systems on the basis of their sensitivity to acetylcholine, adrenaline, or other neurotransmitters; that is, according to whether they are cholinergic, adrenergic, dopaminergic, or otherwise chemically specific. This has proved advantageous in relation to the selective stimulation of a particular system, as employed in studies of the control of ovulation (chapter 10), and eating and drinking (chapter 15).

Study of the action of hormones on nerve cells is of immense importance from the neuroendocrine point of view, for this provides the basis of the feedback action of many factors. As outlined in chapters 7, 10, and 16, work is in progress in analysing the effects of steroid hormones on the electrical activity of single neurones. Studies of the changing firing rate of single cells will not necessarily assist in the understanding of the mode of action of a steroid, although the very fact that the rate of discharge changes is of great significance, but they provide essential information on the locus of action of hormones. The work with hypothalamic islands makes it quite certain that hormones exert a direct action on this part of the brain, which may be quite distinct from any indirect influence derived from an action elsewhere. And in this connection the physiological significance of the phenomenon of the feedback action of pituitary hormones on the brain, the so-called short feedback loop, still remains to be assessed.

There is little doubt that much more effort will be devoted to investigating the extra-hypothalamic control of endocrine function than has hitherto been the case. Now that the working of the hypothalamus is somewhat better understood, meaningful experiments on the effects of lesions, or of stimulation, of the components of the limbic system can be performed. The realization that the amygdala and hippocampus can be antagonistic, and can on occasion inhibit the hypothalamus, will prove to be of immense help in this regard. Improvements will be necessary in the techniques applied to this problem because of the anatomical layout of the amygdala and hippocampus, for it is extraordinarily difficult to ablate the amygdaloid nuclei in a reproducible fashion without involving the hippocampus, or to destroy the hippocampus without a great deal of ancillary brain damage.

Stimulation of the brain can be made very precise in terms of the volume of tissue and the parameters applied, but the selection of the optimal stimulus for any particular structure or system involves much trial and error. Long-term experiments with conscious animals can be simplified by training them to operate a switch controlling the stimulator, and allowing them to determine how long stimulation should continue. This procedure is only effective in those cases when the subject finds the stimulus pleasurable, and only punctuate excitation of any system can be provided, but the results can be informative. It has been found that high levels of progesterone enhance the rate of self-stimulation in rats and rabbits soon after administration (Campbell, 1968), but the effect becomes reversed with continued treatment—a response which recalls the initial facilitation and subsequent inhibition of ovulation by this steroid in the rabbit. Oestrogen appears to have little effect upon the rate of self-stimulation, which does not change in phase with the oestrous cycle, but androgen is facilitatory in the male. Electrical self-stimulation in monkeys, through electrodes in the lateral preoptic region, has been shown to increase the activity of both the adrenal medulla and cortex (McHugh, Black, and Mason, 1966). These endocrine changes closely resembled those occurring with emotional arousal and indicate that the animals were exciting themselves.

An alternative means of eliciting the release of a particular hormone at will is to condition the response. The ovulation occasionally evoked in an oestrous female rabbit upon seeing a buck may provide a familiar example, but subtler responses can be elicited by conditioning procedures, such as changes in blood glucose concentration which match those previously produced by a series of injections of glucagon. Reports of comparable changes in the secretion of other hormones can be anticipated.

It is perhaps inevitable that progress in neuroendocrinology, just as in other fields of endocrine research, marches in step with advances in the assay of hormones. Because of poor sensitivity, it often remains difficult to collect blood samples of sufficient size to allow serial assays of the content of a particular hormone to be performed on one individual. It is, at present, difficult to avoid removing such large amounts of blood as to disturb the equilibrium levels of hormone output and so trigger off compensatory mechanisms. The simultaneous measurement of several hormones in a single specimen of blood or tissue is impracticable at present, in spite of the many advantages in the correlation of the release of several hormones that such a procedure would bring, but will not be long delayed. In those instances where facilities for the assay of several hormones are available in a single laboratory complex, a cooperative study of the effects of brain stimulation upon endocrine activity would yield great dividends. Such an investigation would need to be carefully planned, rigorously performed, and would be laborious, but much information on the neuroendocrine potentialities of brain structures would accrue and many discrepancies in the literature that have arisen because of differing techniques of stimulation or assay, would be resolved. Progress in this direction has been made by Mason and his colleagues, who studied the excretion of 17-hydroxycorticosteroids, adrenaline, noradrenaline, testosterone, and oestrogens, as well as the plasma levels of butanol-extractable iodine, growth hormone, and insulin, in monkeys subjected to periods of conditioned avoidance to electric shocks applied to the feet (Mason, 1968c). The hormonal changes observed fell into two groups, for the levels of the 17-hydroxycorticosteroids, adrenaline, noradrenaline, iodine, and growth hormone all rose initially, while the output of oestrogen, testosterone, and insulin fell. This suggested that the multiple hormonal responses to avoidance may be organized on some overall basis, and Mason (1968c) considers that the hormones of the first group broadly exert 'catabolic' effects on energy metabolism, whereas those in the second group are largely 'anabolic'. However, while indicative of the results that can accrue from an approach of this kind, the concept requires further development and evaluation. The great advances in understanding of the factors concerned in controlling the secretion of growth hormone (chapter 9) have sprung largely from the application of immunoassay procedures, originally developed for the measurement of insulin, to the protein hormones of the pars distalis. Since adequate amounts of pure human growth hormone were available for use in the development of the technique, work on this pituitary factor was undertaken first, but now it is possible to assay other pituitary hormones in this way. Because of the ready clinical application of sensitive hormone assays, and because the hormones from human hypophyses are more easily obtainable in pure form than those of laboratory animals, more is known about the day to day fluctuations in hormone secretion in man than in the usual experimental animals. However, in a number of instances adequate evidence of a direct correlation between endocrine and immunological activity has still to be presented. So far as the steroid hormones are concerned, the parallel and competitive development of assays employing gas chromatography or protein binding is already providing information which is revolutionizing generally accepted concepts, as exemplified by the endocrinology of the menstrual cycle (chapter 10). It is far too early to guess which assay techniques will achieve dominance, for each possesses certain advantages, although for speed and convenience, protein binding methods offer the most promise. However, problems of specificity remain to be overcome.

It is unlikely that every humoral agent at work in the mammal has yet been discovered. Work on the substance produced by the uterus that depresses the secretion of progestins by the corpus luteum provides an excellent case in point, for while the existence of a luteolytic factor of uterine origin is now accepted, the identity of the active agent has yet to be determined. Some evidence is available in support of the possibility that a prostaglandin, one of a class of naturally occurring and highly pharmacologically active fatty acids, causes luteolysis, although definitive proof of this suggestion has still to be provided. The prostaglandins might be of immense physiological significance in other ways (Horton, 1969), and once this becomes established, study of this group of presumptive hormones can begin.

The releasing factors form another group of humoral substances of great interest. Currently, most research effort is devoted to the identification and synthesis of these neurohumoral agents, for they could be of therapeutic value in cases of hypopituitarism. Once this is achieved, attention can be turned to devising and preparing compounds for use as releasing factor antagonists. A substance that prevented CRF from causing ACTH secretion would be a most useful tool in the study of adrenal function, for the result would be equivalent to inhibition of the release of a single pituitary hormone, or a selective 'hormonectomy', without the disadvantages of lack of other pituitary hormones and the general depression of endocrine function that follows surgical ablation of the pituitary gland. Similar considerations apply to the other pituitary hormones and such problems as the separate identity of FSH and LH, and of growth hormone and prolactin in man, would move nearer resolution—always provided that sufficiently selective anti-releasing drugs became available.

There is little doubt that the mammalian body houses endocrine organs that still await discovery. For example, the activities of the subcommissural and subfornical organs referred to in chapter 15 are poorly understood, while suggestions that the salivary glands may act in an endocrine, as well as in an exocrine, fashion have been made. Some critics would include the pineal gland and pars intermedia in this category, on the basis that the relationship between the pineal gland and gonadal function (chapter 14) remains unproven, and that the secretion of the intermediate lobe is of no physiological value. However, interest in the pineal gland is currently in a state of rearousal and seems likely to elicit information of enduring value, while it is improbable that studies on the pars intermedia will stop at analyses of the processes involved in the control of this organ. Comparative studies, or research employing less-common species, will in all likelihood yield the significant facts that will force a reappraisal of present views of the physiology of these two glands. Sir Alan Parkes (1966) has remarked that:

'Of the thousands of different species of mammals, we have substantial knowledge of the reproductive processes of fewer than 100. Nor can we assume that the known types give us a fair cross-section of the world of mammals. Almost every new species which is investigated presents something new and significant. There is no limit to the variations which are being found in what was once thought to be the typical pattern of reproductive cycle. I sometimes wonder how our views on the physiology of reproduction in mammals would have developed had they been based not on domestic and laboratory animals but on seals, bats, elephants, and the like.'

A pointer in this direction comes from the work on the relationship between blinding, the pineal gland, and reproduction in the hamster, which has proved to be more convincing than much of the work on the rat (chapter 14). It is entirely reasonable to suppose that the secretions of the pineal gland and pars intermedia are used to greater effect by species outside the range encountered in a typical laboratory.

The umbrella of neuroendocrinology is likely to be extended to hormones currently considered to be free of guidance by the nervous system. The control of insulin secretion may be a case in point, for Frohman (1969) quotes studies indicating that stimulation of the vagus nerves causes the immediate release of insulin. However, insulin release is not prolonged with continued stimulation and the plasma levels of hormone tend to return to basal levels. The effect of vagal stimulation is blocked by prior treatment with atropine, and cholinergic nerve fibres have been demonstrated around the islet cells in the pancreas. Both adrenaline and noradrenaline have been reported to block the release of insulin from the islet cells, which is induced by glucose or by glucagon, so that the level of insulin secretion could be determined by the prevailing balance between sympathetic and parasympathetic activity. Further, electrolytic destruction of the ventromedial nuclei in the hypothalamus of weanling rats caused hyperinsulinaemia, hyperlipaemia, and excessive fat deposition despite normoglycaemia and an absence of increased food intake or excessive weight gain. Linear growth was also impaired and the animals suffered from an insufficiency of growth hormone. Nevertheless, the fact remains that the transplanted, and so denervated, pancreas continues to secrete insulin and can maintain a normal blood sugar level for weeks.

The striking, and lasting, effects of disturbance of the normal hormonal equilibrium early in life have been discussed in chapter 16, and this field of study is certain to be cultivated intensively. It is likely that the ontogenesis of the neural control of endocrine function will be traced back into foetal life and not equated simply with the formation of the plexus of portal vessels between the hypothalamus and hypophysis. This is foreshadowed by the work of Jost

(1966) on rats, which indicates that the brain is involved in the control of the foetal adrenal gland before the development of the portal vessels. Both decapitation or encephalectomy (which removes the brain while leaving the pituitary gland in place) of the foetus on day 17 slow adrenal growth, implying that the pituitary gland is already under diencephalic influence. For if it were not, the spared hypophysis should be able to support adrenal enlargement. This conclusion is substantiated by the observation that the injection of extracts of hypothalamus, or of CRF, caused adrenal growth in encephalectomized, but not in decapitated, animals. Further, the injection of formalin into foetuses caused adrenal ascorbic acid depletion in intact but not in encephalectomized individuals. There is also a delay in the maturation of the thyroid gland after decapitation, but the thyroid is less indirectly dependent upon the hypothalamus than the adrenal.

The action of hormones on the brain is likely to attract increasing attention, not only because of the effects produced upon behaviour and endocrine function, and referred to earlier, but because it is becoming possible to trace the locus of action with ever greater precision with the aid of radioactive-tracer labelled compounds. Study of the mechanism of action of hormones on the brain lies in the realm of neurochemistry but the way in which they guide the laying-down of the foundations of subsequent behavioural patterns and control mechanisms, provides a particularly fertile field of neuroendocrine investigation. Greater emphasis will no doubt be placed upon the integration of neuroendocrine mechanisms within the hypothalamus, and indeed other parts of the brain. The means by which simultaneous control of six anterior pituitary hormones is exercised by a few cubic millimetres of neural tissue poses a perplexing problem, particularly when the high degree of overlap in the variety of stimuli affecting a single hypothalamic neurone is called to mind. It is possible that non-specificity on the part of many hypothalamic neurones provides the basis of an integrative mechanism, for the eventual discharge rate of each cell represents the outcome of a variety of contending influences. Some common denominator may be utilized in guiding several processes, as in the case of the blood level of glucose which provides the signal for modulation of the secretion of adrenaline by the adrenal medulla, the secretion of growth hormone, and the regulation of food intake. Brobeck (1960), for example, has argued that the hypothalamus might act as an integrator for the four variables of energy balance, which are food intake, activity, body temperature, and body weight, and that body temperature, or some signal derived from it, is an important factor in the control of food intake. This view has not won general acceptance, but it has been observed in rats that an increase in body temperature was followed by an increase in the electrical activity of the 'satiety centre' in the hypothalamus, and a decrease in the activity of the 'feeding centre'. Slight increases in the frequency of the response recorded in other, control, brain areas also occurred with a rise in body temperature, so that the decrease in mean frequency of activity in the lateral hypothalamus seems significant (Traylor and Blackburn, 1969). It is evident that, besides their reaction to changes in the blood glucose level, single units in the ventromedial nuclei of the hypothalamus respond to a wide variety of environmental stimuli. This broad spectrum of afferent input led Campbell, Bindra, Krebs, and Ferenchak (1969) to postulate that lesions in these nuclei may produce hyperphagia by making the animal less distractable to novel environmental stimuli. Conversely, excitation of the nuclei would make the animals pay less attention to food, for they would be more readily distracted.

Much needs to be learned of the activities of the limbic system in controlling endocrine function, for it is probably through this region of the brain that emotions most readily influence hormonal secretion.

The fact that the action of hormones in the rat, and man, may not be typical of the mammalia as a whole, also bears repetition, and this is well illustrated by prolactin, which, as described in chapter 11, supports corpus luteum activity in the rat, mouse, and ferret, but is ineffective in the rabbit, guinea-pig, cow, or woman. And it is not yet certain whether the human hypophysis secretes prolactin and growth hormone, or whether one secretory product suffices to cover the functions of both hormones. To some degree, species specificity in the action of protein hormones may, on occasion, account for the lack of response to an extract prepared from the gland of a different species, but too many examples of highly sensitive responses are available for this explanation to be fully satisfactory. Nevertheless, it is a measure of the restrictions placed on research of this kind that only recently have samples of purified rat pituitary hormones become available for research purposes, and then particularly in relation to the immunoassay of these factors. The reason for this shortcoming lies more in the realm of the financing of research, than in the lack of rat pituitary glands.

Alongside the acquisition of knowledge about endocrine function in common laboratory animals and in man, the intensive study of more exotic species is gradually gaining momentum in an effort to account for the great variety of patterns of hormonal and reproductive activity. It is perhaps over-easy to account for the sex cycle in the rat on the basis of daily fluctuations in hypothalamic activity, with a regular surge or drive toward gonadotrophin secretion, although models of the system remain somewhat intricate. In other forms with an oestrous, or menstrual, cycle of greater length the rhythmic occurrence of ovulation is less readily explained (chapter 10), and it is to be hoped that the genesis of rhythms will be studied at greater depth.

References

ABRAHAMS, V. C., LANGWORTH, E. P., & THEOBALD, G. W. (1964). Potentials evoked in the hypothalamus and cerebral cortex by electrical stimulation of the uterus. *Nature, Lond.,* **203,** 654–656.

ACHER, R. (1966). Chemistry of neurohypophysial hormones. In: *The Pituitary Gland.* Eds. G. W. Harris & B. T. Donovan. **3,** 269–287. London: Butterworths.

ACHER, R. (1968). Neurophysin and neurohypophysial hormones. *Proc. Roy. Soc. B.,* **170,** 7–16.

ADER, R. (1969). Early experiences accelerate maturation of the 24-hour adreno-cortical rhythm. *Science,* **163,** 1225–1226.

ADOLPH, E. F. (1961). Early concepts of physiological regulations. *Physiol. Rev.,* **41,** 737–770.

ALTMAN, J., DAS, G. D., & ANDERSON, W. J. (1968). Effects of infantile handling on morphological development of the rat brain: an exploratory study. *Develop. Psychobiol.,* **1,** 10–20.

AMES, R. G., MOORE, D. H., & van DYKE, H. B. (1950). The excretion of posterior pituitary antidiuretic hormone in the urine and its detection in the blood. *Endocrinology*, **46**, 215–227.

ANAND, B. K. (1961). Nervous regulation of food intake. *Physiol. Rev.*, **41**, 677–708.

ANAND, B. K. & BROBECK, J. R. (1951). Localization of a 'feeding center' in the hypothalamus of the rat. *Proc. Soc. exper. Biol. Med.*, **77**, 323–324.

ANAND, B. K., CHHINA, G. S., SHARMA, K. N., DUA, S., & SINGH, B. (1964). Activity of single neurons in the hypothalamic feeding centers: effect of glucose. *Amer. J. Physiol.*, **207**, 1146–1154.

ANAND, B. K., DUA, S., & SHOENBERG, K. (1955). Hypothalamic control of food intake in cats and monkeys. *J. Physiol., Lond.*, **127**, 143–152.

ANDERSSON, B. (1953). The effect of injections of hypertonic NaCl-solutions into different parts of the hypothalamus of goats. *Acta physiol. scand.*, **28**, 188–201.

ANDERSSON, B. (1964). Hypothalamic temperature and thyroid activity. In: *Brain-Thyroid Relationships*. Eds. M. P. Cameron & M. O'Connor. Ciba Foundation Study Group, No. 18, 35–50. London: Churchill.

ANDERSSON, B. (1966). The physiology of thirst. In: *Progress in Physiological Psychology*. Eds. E. Stellar & J. M. Sprague. **1**, 191–207. New York: Academic Press.

ANDERSSON, B., GALE, C. C., HÖKFELT, B., & LARSSON, B. (1965). Acute and chronic effects of preoptic lesions. *Acta physiol. scand.*, **65**, 45–60.

ANDERSSON, B. & McCANN, S. M. (1955). Drinking, antidiuresis, and milk ejection from electrical stimulation within the hypothalamus of the goat. *Acta. physiol. scand.*, **35**, 191–201.

ASCHOFF, J. (1965). Circadian rhythms in man. *Science*, **148**, 1427–1432.

ASHMORE, J. & MORGAN, D. (1967). Metabolic effects of adrenal glucocorticoid hormones: carbohydrate, protein, lipid, and nucleic acid metabolism. In: *The Adrenal Cortex*. Ed. A. B. Eisenstein. 249–267. London: Churchill.

AULSEBROOK, L. H. & HOLLAND, R. C. (1969a). Central regulation of oxytocin release with and without vasopressin release. *Amer. J. Physiol.*, **216**, 818–829.

AULSEBROOK, L. H. & HOLLAND, R. C. (1969b). Central inhibition of oxytocin release. *Amer. J. Physiol.*, **216**, 830–842.

AVERILL, R. L. W. (1969). Depression of thyrotropin releasing factor induction of thyrotropin release by thyroxine in small doses. *Endocrinology*, **85**, 67–71.

AVERILL, R. L. W., SALAMAN, D. F., & WORTHINGTON, W. C. (1966). Thyrotropin releasing factor in hypophyseal portal blood. *Nature, Lond.*, **211**, 144–145.

BAILE, C. A. (1968). Regulation of feed intake in ruminants. *Fed. Proc.*, **27**, 1361–1366.

BARD, P. (1940). The hypothalamus and sexual behavior. *Res. Pub. Ass. nerv. ment. Dis.*, **20**, 551–576.

BARGMANN, W. (1966). Neurosecretion. *Int. Rev. Cytol.*, **19**, 183–201.

BARRACLOUGH, C. A. (1968). Alterations in reproductive function following prenatal and early postnatal exposure to hormones. In: *Advances in Reproductive Physiology*. Ed. A. McLaren. **3**, 81–112. London: Logos Press.

BATES, R. W. & CONDLIFFE, P. G. (1966). The physiology and chemistry of thyroid stimulating hormone. In: *The Pituitary Gland*. Eds. G. W. Harris & B. T. Donovan. **1**, 374–410. London: Butterworths.

BAXTER, B. L. (1967). Comparison of the behavioral effects of electrical or chemical stimulation applied at the same brain loci. *Exper. Neurol.*, **19**, 412–432.

BAXTER, B. L. (1968). Elicitation of emotional behavior by electrical or chemical stimulation applied at the same loci in cat mesencephalon. *Exper. Neurol.*, **21**, 1–10.

BAYLISS, W. M. & STARLING, E. H. (1902). See Brooks, Gilbert, Levey, & Curtis, 1962.

BEDDOW, D. G. & McCANN, S. M. (1969). Effect of median eminence lesions on the function of multiple pituitary homografts with particular reference to the secretion of gonadotrophins and growth hormone. *Endocrinology*, **84**, 595–605.

BELESLIN, D., BISSETT, G. W., HALDAR, J., & POLAK, R. L. (1967). The release of vasopressin without oxytocin in response to haemorrhage. *Proc. Roy. Soc. B.*, **166**, 443–458.

BENOIT, J. (1964). The role of the eye and of the hypothalamus in the photostimulation of gonads in the duck. *Ann. N.Y. Acad. Sci.*, **117**, 204–215.

BERN, H. A. (1966). On the production of hormones by neurones and the role of neurosecretion in neuroendocrine mechanisms. *Sympos. Soc. Experim. Biol.*, **20**, 325–344.

BERTHOLD, A. A. (1849). See Brooks, Gilbert, Levey, & Curtis, 1962.

van BEUGEN, L. & van der WERFF ten BOSCH, J. J. (1961). Rat thyroid activity and cold response after removal of frontal parts of the brain. *Acta endocr., Copenhagen*, **37**, 470–478.

BEYER, C., RAMIREZ, V. D., WHITMOYER, D. I., & SAWYER, C. H. (1967). Effects of hormones on the electrical activity of the brain in the rat and rabbit. *Exper. Neurol.*, **18**, 313–326.

BEYER, C. & SAWYER, C. H. (1969). Hypothalamic unit activity related to control of the pituitary gland. In: *Frontiers in Neuroendocrinology, 1969*. Eds. W. F. Ganong & L. Martini. 255–287. New York: Oxford University Press.

BIDDER, T. G. (1968). Modification of progesterone translocation into brain. *Endocrinology*, **83**, 1353–1355.

BISSETT, G. W., HILTON, S. M., & POISNER, A. M. (1967). Hypothalamic pathways for independent release of vasopressin and oxytocin. *Proc. Roy. Soc. B.*, **166**, 422–442.

BJÖRKLUND, A., ENEMAR, A., & FALCK, B. (1968). Monoamines in the hypothalamo-hypophyseal system of the mouse with special reference to the ontogenetic aspects. *Z. Zellforsch.*, **89**, 590–607.

BLAND, K. P. & DONOVAN, B. T. (1966). The uterus and the control of ovarian function. In: *Advances in Reproductive Physiology*. Ed. A. McLaren. **1**, 179–214. London: Logos Press.

BOGDANOVE, E. M. (1963). Direct gonad-pituitary feedback: an analysis of effects of intracranial estrogenic depots on gonadotrophin secretion. *Endocrinology*, **73**, 696–712.

BOHUS, B., NYAKAS, C., & LISSÁK, K. (1968). Involvement of suprahypothalamic structures in the hormonal feedback action of corticosteroids. *Acta. physiol. Hung.*, **34**, 1–8.

BØLER, J., ENZMANN, F., FOLKERS, K., BOWERS, C. Y., & SCHALLY, A. V. (1969). The identity of chemical and hormonal properties of the thyrotropin releasing hormone and pyroglutamyl-histidyl-proline amide. *Biochem. Biophys. Res. Comm.*, **37**, 705–710.

BOURGERY (1845). Quoted by Green, 1966b.

BOYD, R. & JOHNSON, D. C. (1968). Gonadotrophin patterns in male and female rats: inhibition of LH release by testosterone propionate in animals gonadectomized at puberty. *Acta endocr. Copenhagen*, **58**, 600–612.

BREGGIN, P. R. (1964). The psychophysiology of anxiety. *J. Nerv. Ment. Dis.*, **139**, 558–568.

BROBECK, J. R. (1946). Mechanism of the development of obesity in animals with hypothalamic lesions. *Physiol. Rev.*, **26**, 541–559.

BROBECK, J. R. (1960). Food and temperature. *Recent Progr. Horm. Res.*, **16**, 439–459.

BRODISH, A. (1963). Diffuse hypothalamic system for the regulation of ACTH secretion. *Endocrinology*, **73**, 727–735.

BROOKS, C. McC., GILBERT, J. L., LEVEY, H. A., & CURTIS, D. R. (1962). *Humors, Hormones, and Neurosecretions*. State University of New York.

BROWN, J. J., FRASER, R., LEVER, A. F., & ROBERTSON, J. I. S. (1968). Renin and angiotensin in the control of water and electrolyte balance; relation to aldosterone. In: *Recent Advances in Endocrinology*. 8th Ed. Ed. V. H. T. James. 271–292. London: Churchill.

BROWN-GRANT, K. (1966). The control of TSH secretion. In: *The Pituitary Gland*. Eds. G. W. Harris & B. T. Donovan, **2**, 235–269. London: Butterworths.

BROWN-GRANT, K. (1967). The control of thyroid secretion. *J. Clin. path.*, **20**. Suppl. 1, 327–332.

BRUCE, H. M. (1970). Pheromones. *Brit. med. Bull.*, **26**, 10–13.

186

BURGUS, R., DUNN, T. F., DESIDERIO, D., & GUILLEMIN, R. (1969). Structure moléculaire du facteur hypothalamique hypophysiotrope TRF d'origine ovine: mise en evidence par spectrométrie de masse de la séquence PCA-His-Pro-NH$_2$. *C. R. Acad. Sci. Paris*, **269**, 1870–1873.

BURGUS, R., DUNN, T. F., WARD, D. N., VALE, W., AMOSS, M., & GUILLEMIN, R. (1969). Dérivés polypeptidiques de synthèse doués d'activité hypophysiotrope TRF. *C. R. Acad. Sci. Paris*, **268**, 2116–2118.

BUSE, M. G., FULMER, J. D., KANSAL, P. C., & WORTHINGTON, W. C. (1970). The effects of hypophysial portal plasma on the content of immunoreactive growth hormone in the anterior pituitary. *J. Physiol. Lond.*, **206**, 243–256.

BUTCHER, L. L. & FOX, S. S. (1968). Motor effects of copper in the caudate nucleus: reversible lesions with ion-exchange resin beads. *Science*, **160**, 1237–1239.

BUTLER, J. E. M. & DONOVAN, B. T. (1969). Consequences of hypothalamic de-afferentation in female rats and guinea-pigs. *J. Endocrinol.*, **43**, xx–xxi.

CAMPBELL, J. F., BINDRA, D., KREBS, H., & FERENCHAK, R. P. (1969). Responses of single units of the hypothalamic ventromedial nucleus to environmental stimuli. *Physiol. Behav.*, **4**, 183–187.

CAMPBELL, H. J. (1968). Acute effects of pregnene steroids on septal self-stimulation in the rabbit. *J. Physiol. Lond.*, **196**, 134P–135P.

CAMPBELL, H. J. & EAYRS, J. T. (1965). Influence of hormones on the central nervous system. *Brit. med. Bull.*, **21**, 81–86.

CAMPBELL, H. J., FEUER, G., & HARRIS, G. W. (1964). The effect of intrapituitary infusion of median eminence and other brain extracts on anterior pituitary gonado-trophic secretion. *J. Physiol. Lond.*, **170**, 474–486.

CANNON, W. B. & ROSENBLUETH, A. (1937). *Autonomic Neuro-effector Systems*. New York: Macmillan.

CHAU, S. S., FITZPATRICK, R. J., & JAMIESON, B. (1969). Diabetes insipidus and parturition. *J. Obstet. Gynaec. Brit. Cwlth.*, **76**, 444–450.

CHEIFETZ, P., GAFFUD, N., & DINGMAN, J. F. (1968). Effects of bilateral adrenal-ectomy and continuous light on the circadian rhythm of corticotropin in female rats. *Endocrinology*, **82**, 1117–1124.

CHEN, C.-L., VOOGT, J. L., & MEITES, J. (1968). Effect of median eminence implants of FSH, LH or prolactin on luteal function in the rat. *Endocrinology*, **83**, 1273–1277.

CHHINA, G. S. & ANAND, B. K. (1969). Responses of neurones in the hypothalamus and limbic system to genital stimulation in adult and immature monkeys. *Brain Res.*, **13**, 511–521.

CHRISTIAN, J. J. & DAVIES, D. E. (1964). Endocrines, behavior, and population. *Science*, **146**, 1550–1560.

COLE, L. C. (1957). Biological clock in the unicorn. *Science*, **125**, 874–876.

CONNEY, A. H. (1967). Pharmacological implications of microsomal enzyme induc-tion. *Pharm. Rev.*, **19**, 317–366.

COPPOLA, J. A. (1968). The apparent involvement of the sympathetic nervous system in the gonadotrophin secretion of female rats. *J. Reprod. Fert.*, Suppl. 4.

CORSON, S. A. (1966). Conditioning of water and electrolyte excretion. *Proc. Assoc. Res. Nerv. Ment. Dis.*, **43**, 140–198.

COURY, J. N. (1967). Neural correlates of food and water intake in the rat. *Science*, **156**, 1763–1765.

COWIE, A. T. (1969). Variations in the yield and composition of the milk during lactation in the rabbit and the galactopoietic effect of prolactin. *J. Endocrinol.*, **44**, 437–450.

COWIE, A. T., HARTMANN, P. E., & TURVEY, A. (1969). The maintenance of lactation in the rabbit after hypophysectomy. *J. Endocrinol.*, **43**, 651–662.

COX, V. C., KAKOLEWSKI, J. W., & VALENSTEIN, E. S. (1969). Ventromedial hypothalamic lesions and changes in body weight and food consumption in male and female rats. *J. comp. Physiol. Psychol.*, **67**, 320–326.

CRITCHLOW, V. (1963). The role of light in the neuroendocrine system. In: *Advances in Neuroendocrinology.* Ed. A. V. Nalbandov. 377–402. Urbana: University of Illinois Press.

CRITCHLOW, V. & BAR-SELA, M. E. (1967). Control of the onset of puberty. In: *Neuroendocrinology.* Eds. L. Martini & W. F. Ganong. 2, 101–162. New York: Academic Press.

CRONE, C. & SILVER, I. A. (1964). Quoted by Cross, 1964.

CROSBY, E. C., HUMPHREY, T., & LAUER, E. W. (1962). *Correlative Anatomy of the Nervous System.* New York: Macmillan.

CROSS, B. A. (1964). The hypothalamus in mammalian homeostasis. *Symp. Soc. exper. Biol.,* **18,** 157–193.

CROSS, B. A. (1966). Neural control of oxytocin secretion. In: *Neuroendocrinology.* Eds. L. Martini & W. F. Ganong. **1,** 217–259. New York: Academic Press.

CROSS, B. A. & KITAY, J. I. (1967). Unit activity in diencephalic islands. *Exper. Neurol.,* **19,** 316–330.

CROSS, B. A. & SILVER, I. A. (1966). Electrophysiological studies on the hypothalamus. *Brit. med. Bull.,* **22,** 254–260.

CROW, L. T. (1964). Subcommissural organ, lateral hypothalamus and dorsal longitudinal fasciculus in water and salt metabolism. In: *Thirst,* Ed. M. J. Wayner, 473–481. Oxford: Pergamon Press.

CROWE, S. J., CUSHING, H., & HOMANS, J. (1912). See Brooks, Gilbert, Levey, & Curtis, 1962.

CSAPO, A. I. & WOOD, C. (1968). The endocrine control of the initiation of labour in the human. In: *Recent Advances in Endocrinology.* 8th Edn. Ed. V. H. T. James. 207–239. London: Churchill.

DAUGHADAY, W. H., PEAKE, G. T., BIRGE, C. A., & MARIZ, I. K. (1968). The influence of endocrine factors on the concentration of growth hormone in rat pituitary. *Proc. 1st Intern. Sympos. Growth Hormone.* Excerpta Medica Congress Series, No. 158, 238–252.

DAVIDSON, J. M. (1967). Neuroendocrine mechanisms in the control of spermatogenesis. *J. Reprod. Fert.,* Suppl. 2, 103–115.

DAVIS, J. D., GALLAGHER, R. L., & LADOVE, R. (1967). Food intake controlled by a blood factor. *Science,* **156,** 1247–1248.

DAVIS, J. O. (1967). The regulation of aldosterone secretion. In: *The Adrenal Cortex.* Ed. A. B. Eisenstein. 203–247. London: Churchill.

DAVIS, J. O., CARPENTER, C. C. J., AYERS, C. R., HOLMAN, J. E., & BAHN, R. C. (1961). Evidence for secretion of an aldosterone-stimulating hormone by the kidney. *J. clin. Invest.,* **40,** 684–696.

DAVISON, A. N. & DOBBING, J. (1968). The developing brain. In: *Applied Neurochemistry.* Eds. A. N. Davison & J. Dobbing. 253–286. Oxford: Blackwell.

DEBACKERE, M., PEETERS, G., & TUYTTENS, N. (1961). Reflex release of an oxytocic hormone by stimulation of genital organs in male and female sheep studied by a cross-circulation technique. *J. Endocrinol.,* **22,** 321–334.

DE COURSEY, P. (1964). Function of a light response rhythm in hamsters. *J. Cell. Comp. Physiol.,* **63,** 189–196.

DENEF, C. & DE MOOR, P. (1968). The 'puberty' of the rat liver. II. Permanent changes in steroid metabolizing enzymes after treatment with a single injection of testosterone propionate at birth. *Endocrinology,* **83,** 791–798.

DE ROBERTIS, E. D. P. (1964). *Histophysiology of Synapses and Neurosecretion.* Oxford: Pergamon Press.

DEUBEN, R. R. & MEITES, J. (1964). Stimulation of pituitary growth hormone release by a hypothalamic extract *in vitro. Endocrinology,* **74,** 408–414.

DEWHURST, K. E., EL KABIR, D. J., HARRIS, G. W., & MANDELBROTE, B. M. (1968). A review of the effect of stress on the activity of the central nervous-pituitary-thyroid axis in animals and man. *Confin. Neurol.,* **30,** 161–196.

DOBBING, J. (1961). The blood-brain barrier. *Physiol. Rev.,* **41,** 130–188.

DOBBING, J. (1968). The blood-brain barrier. In: *Applied Neurochemistry.* Eds. A. N. Davison & J. Dobbing. 317–331. Oxford: Blackwell.

DÖCKE, F., DORNER, G., & VOIGT, K.-H. (1968). A possible mechanism of the ovulation-inhibiting effect of chlormadinone acetate in the rat. *J. Endocrinol.,* **41,** 353–362.

DONOVAN, B. T. (1963). The timing of puberty. *Scientific Basis of Medicine Annual Reviews,* 53–75.

DONOVAN, B. T. (1966). The regulation of the secretion of follicle-stimulating hormone. In: *The Pituitary Gland.* Eds. G. W. Harris & B. T. Donovan. **2,** 49–98. London: Butterworths.

DONOVAN, B. T. (1967). Existence of a luteolytic hormone in the uterus of the guinea-pig. In: *Reproduction in the Female Mammal.* Eds. G. E. Lamming & E. C. Amoroso. 317–337. London: Butterworths.

DONOVAN, B. T. & HARRIS, G. W. (1955). Neurohumoral mechanisms in reproduction. *Brit. med. Bull.,* **11,** 93–97.

DONOVAN, B. T. & HARRIS, G. W. (1966). Neurohumoral mechanisms in reproduction. In: *Marshall's Physiology of Reproduction.* Ed. A. S. Parkes. **3,** 301–378. London: Longmans Green.

DONOVAN, B. T. & LOCKHART, A. N. (1970). The brain and gonadal function in health and disease. In: *Modern Trends in Endocrinology.* Eds. H. Gardiner-Hill & F. T. G. Prunty. In press. London: Butterworths.

DONOVAN, B. T. & van der WERFF ten BOSCH, J. J. (1957). The hypothalamus and lactation in the rabbit. *J. Physiol. Lond.,* **137,** 410–420.

DONOVAN, B. T. & van der WERFF ten BOSCH, J. J. (1959a). The relationship of the hypothalamus to oestrus in the ferret. *J. Physiol. Lond.,* **147,** 93–108.

DONOVAN, B. T. & van der WERFF ten BOSCH, J. J. (1959b). The hypothalamus and sexual maturation in the rat. *J. Physiol. Lond.,* **147,** 78–92.

DONOVAN, B. T. & van der WERFF ten BOSCH, J. J. (1965). *Physiology of puberty.* London: Arnold.

DÖRNER, G. & STAUDT, J. (1968). Structural changes in the preoptic anterior hypothalamic area of the male rat, following neonatal castration and androgen substitution. *Neuroendocrinology,* **3,** 136–140.

DUNÉR, H. (1953). The influence of the blood glucose level on the secretion of adrenaline and noradrenaline from the suprarenal. *Acta physiol. scand.,* **28,** Suppl. 102.

DUNN, J. & CRITCHLOW, V. (1969). Feedback suppression of pituitary-adrenal function in rats with pituitary islands. *Life Sciences,* **8,** 9–16.

EAYRS, J. T. (1968). Developmental relationships between brain and thyroid. In: *Endocrinology and Human Behaviour.* Ed. R. P. Michael. 239–253. London: Oxford University Press.

EDSTRÖM, R. (1964). Recent developments of the blood-brain barrier concept. *Int. Rev. Neurobiol.,* **7,** 153–190.

EGDAHL, R. H. (1968). Excitation and inhibition of ACTH secretion. *Mem. Soc. Endocrinol.,* **17,** 29–37.

EPSTEIN, A. N. & TEITELBAUM, P. (1967). Specific loss of the hypoglycemic control of feeding in recovered lateral rats. *Amer. J. Physiol.,* **213,** 1159–1167.

ETKIN, W. (1967). Relation of the pars intermedia to the hypothalamus. In: *Neuroendocrinology.* Eds. L. Martini & W. F. Ganong. **2,** 261–282. New York: Academic Press.

von EULER, U. S. (1967). Adrenal medullary secretion and its neural control. In: *Neuroendocrinology.* Eds. L. Martini & W. F. Ganong. **2,** 283–333. New York: Academic Press.

EVANS, H. M., SPARKS, L. L., & DIXON, J. S. (1966). The physiology and chemistry of adrenocorticotrophin. In: *The Pituitary Gland.* Eds. G. W. Harris & B. T. Donovan. **2,** 317–373. London: Butterworths.

EVANS, J. S. & NIKITOVITCH-WINER, M. B. (1969). Functional reactivation and cytological restoration of pituitary grafts by continuous local intravascular infusion of median eminence extracts. *Neuroendocrinology,* **4,** 83–100.

EVERETT, J. W. (1964). Central neural control of reproductive functions of the adenohypophysis. *Physiol. Rev.*, **44**, 373–431.

EVERETT, J. W. (1966). The control of the secretion of prolactin. In: *The Pituitary Gland*. Eds. G. W. Harris & B. T. Donovan. **2**, 166–194. London: Butterworths.

EVERETT, J. W. (1968). 'Delayed pseudopregnancy' in the rat, a tool for the study of central neural mechanisms in reproduction. In: *Reproduction and Sexual Behavior*. Ed. M. Diamond. 25–31. Indiana University Press.

EVERETT, J. W. & QUINN, D. L. (1966). Differential hypothalamic mechanisms inciting ovulation and pseudopregnancy in the rat. *Endocrinology*, **78**, 141–150.

EXLEY, D. & CORKER, C. S. (1966). The human male cycle of urinary oestrone and 17-oxosteroids. *J. Endocrinol.*, **35**, 83–99.

FABIAN, M., FORSLING, M. L., JONES, J. J., & LEE, J. (1969). The release, clearance and plasma protein binding of oxytocin in the anaesthetized rat. *J. Endocrinol.*, **43**, 175–189.

FALCK, B. (1962). Observations on the possibilities of the cellular localization of monoamines by a fluorescence method. *Acta physiol. scand.*, **56**, Suppl. 197.

FALCK, B. & OWMAN, C. (1965). A detailed methodological description of the fluorescence method for the cellular demonstration of biogenic amines. *Acta Univ. Lund.*, Sect. II, No. 7.

FARNER, D. S. (1965). Circadian systems in the photoperiodic responses of vertebrates. In: *Circadian Clocks*, 357–369. Amsterdam: N. Holland.

FELDMAN, S., CONFORTI, N., CHOWERS, I., & DAVIDSON, J. M. (1968). Differential effects of hypothalamic deafferentation on responses to different stresses. *Israel J. Med. Sci.*, **4**, 908–910.

FERGUSON, J. K. W. (1941). A study of the motility of the intact uterus at term. *Surg. Gynec. Obstet.*, **73**, 359–366.

FERIN, M., TEMPONE, A., ZIMMERING, P. E., & VANDE WIELE, R. L. (1969). Effect of antibodies to 17β-estradiol and progesterone on the estrous cycle of the rat. *Endocrinology*, **85**, 1070–1078.

FISHER, A. E. & COURY, J. N. (1964). Chemical tracing of neural pathways mediating the thirst drive. In: *Thirst*. Ed. M. J. Wayner. 515–529. Oxford: Pergamon Press.

FISHER, C., INGRAM, W. R., & RANSON, S. W. (1938). *Diabetes Insipidus and the Neuro-hormonal Control of Water Balance*. Michigan: Edward.

FITZPATRICK, R. J. (1966). The posterior pituitary gland and the female reproductive tract. In: *The Pituitary Gland*. Eds. G. W. Harris & B. T. Donovan. **3**, 453–504. London: Butterworths.

FITZSIMONS, J. T. (1969). The role of a renal thirst factor in drinking induced by extracellular stimuli. *J. Physiol. Lond.*, **201**, 349–368.

FLAMENT-DURAND, J. & DESCLIN, L. (1968). A topographical study of an hypothalamic region with a 'thyrotrophic' action. *J. Endocrinol.*, **41**, 531–539.

FLERKÓ, B., MESS, B., & ILLEI-DONHOFFER, A. (1969). On the mechanism of androgen sterilization. *Neuroendocrinology*, **4**, 164–169.

FORTIER, C. (1966). Nervous control of ACTH secretion. In: *The Pituitary Gland*. Eds. G. W. Harris & B. T. Donovan. **2**, 195–234. London: Butterworths.

FRANCIS, R. P. & MALVEN, P. V. (1968). Duration of pseudopregnancy following intrapituitary implantation of estradiol and progesterone. *Proc. Soc. exper. Biol. Med.*, **129**, 207–210.

FROHMAN, L. A. (1969). The endocrine function of the pancreas. *Ann. Rev. Physiol.*, **31**, 353–382.

FROHMAN, L. A., BERNADIS, L. L., & KANT, K. J. (1968). Hypothalamic stimulation of growth hormone secretion. *Science*, **162**, 580–582.

FUNKENSTEIN, D. H. (1956). Nor-epinephrine-like and epinephrine-like substances in relation to human behavior. *J. Nerv. Ment. Dis.*, **124**, 58–68.

FUXE, K. & HÖKFELT, T. (1969). Catecholamines in the hypothalamus and pituitary gland. In: *Frontiers in Neuroendocrinology, 1969*. Eds. W. F. Ganong & L. Martini. 47–96. New York: Oxford University Press.

GABE, M. (1966). *Neurosecretion*. Oxford: Pergamon Press.

GARCIA, J. F. & GESCHWIND, I. I. (1968). Investigation of growth hormone secretion in selected mammalian species. *Proc. 1st Int. Sympos. Growth Hormone.* Excerpta Med. Congress Series. No. 158, 267–291.

GASTON, S. & MENAKER, M. (1967). Photoperiodic control of hamster testis. *Science,* **158,** 925–928.

GASTON, S. & MENAKER, M. (1968). Pineal function: the biological clock in the sparrow? *Science,* **160,** 1125–1127.

GAUER, O. H. (1968). Osmocontrol versus volume control. *Fed. Proc.,* **27,** 1132–1136.

GAUER, O. H. & HENRY, J. P. (1963). Circulatory basis of fluid volume control. *Physiol. Rev.,* **43,** 423–481.

GIBSON, J. G. (1962). Emotions and the thyroid gland: a critical appraisal. *J. Psychosom. Res.,* **6,** 93–116.

GILBERT, G. J. (1964). The subcommissural organ and water-electrolyte metabolism. In: *Thirst.* Ed. M. J. Wayner. 457–471. Oxford: Pergamon Press.

GINSBURG, M. (1968). Molecular aspects of neurohypophysial hormone release. *Proc. Roy. Soc. B.,* **170,** 27–36.

GLICK, S. M. & GOLDSMITH, S. (1968). The physiology of growth hormone secretion. *Proc. 1st Int. Sympos. Growth Hormone.* Excerpta Medica Congress Series, No. 158, 84–88.

GODDARD, G. V. (1964). Functions of the amygdala. *Psychol. Bull.,* **62,** 89–109.

GOLDFIEN, A. & GANONG, W. F. (1962). Adrenal medullary and adrenal cortical response to stimulation of diencephalon. *Amer. J. Physiol.,* **202,** 205–211.

GOLDMAN, B. D., KAMBERI, I. A., SIITERI, P. K., & PORTER, J. C. (1969). Temporal relationship of progestin secretion, LH release and ovulation in rats. *Endocrinology,* **85,** 1137–1143.

GOODALL, McC. (1951). Studies of adrenaline and noradrenaline in mammalian heart and suprarenals. *Acta physiol. scand.,* **24,** Suppl. 85.

GORSKI, R. A. (1966). Localization and sexual differentiation of the nervous structures which regulate ovulation. *J. Reprod. Fert.,* Suppl. 1, 67–88.

GRANT, L. D. & JARRARD, L. E. (1968). Functional dissociation within hippocampus. *Brain Res.,* **10,** 392–401.

GRANT, N. (1967). Metabolic effects of adrenal glucocorticoid hormones: other effects. In: *The Adrenal Cortex.* Ed. A. B. Eisenstein. 269–292. London: Churchill.

GRAVES, R. J. (1835). see Brooks, Gilbert, Levey, & Curtis, 1962.

GREEN, J. D. (1966a). The comparative anatomy of the portal vascular system and of the innervation of the hypophysis. In: *The Pituitary Gland.* Eds. G. W. Harris & B. T. Donovan. **1,** 127–146. London: Butterworths.

GREEN, J. D. (1966b). Microanatomical aspects of the formation of neurohypophysial hormones and neurosecretion. In: *The Pituitary Gland.* Eds. G. W. Harris & B. T. Donovan. **3,** 240–268. London: Butterworths.

GREEN, J. D., CLEMENTE, C. D., & de GROOT, J. (1957). Rhinencephalic lesions and behavior in cats. *J. comp. Neurol.,* **108,** 505–545.

GREEN, J. D. & HARRIS, G. W. (1947). The neurovascular link between the neurohypophysis and adenohypophysis. *J. Endocrinol.,* **5,** 136–146.

GREEP, R. O. (1966). Some current perspectives and future projections in endocrinology. *Proc. 6th Pan-American Congress of Endocrinology.* Excerpta Medica Congress Series, No. 112, 118–124.

de GROOT, J. (1959). The rat forebrain in stereotaxic coordinates. *Trans. Roy. Nether. Acad. Sci.,* **52,** 1–40.

de GROOT, J. (1962). In discussion. *Proc. XXII Int. Cong. Physiol.,* **1,** 623–624.

de GROOT, J. (1966a). Limbic and other neural pathways that regulate endocrine function. In: *Neuroendocrinology.* Eds. L. Martini & W. F. Ganong. **1,** 81–106. New York: Academic Press.

de GROOT, J. (1966b). Bibliography of stereotaxic and other brain atlases arranged by species. In: *Neuroendocrinology.* Eds. L. Martini & W. F. Ganong. **1,** 133–137. New York: Academic Press.

de GROOT, J. & HARRIS, G. W. (1950). Hypothalamic control of anterior pituitary and blood lymphocytes. *J. Physiol. Lond.*, **111**, 335–346.

GROSS, F. (1968). Control of aldosterone secretion by the renin-angiotensin system and by corticotropin. *Adv. Intern. Med.*, **14**, 281–339.

GROSSMAN, S. P. (1964). Some neurochemical aspects of the central regulation of thirst. In: *Thirst*. Ed. M. J. Wayner. 487–511. Oxford: Pergamon Press.

GROSSMAN, S. P. (1967a). *A Textbook of Physiological Psychology*. New York: Wiley.

GROSSMAN, S. P. (1967b). The central regulation of food and water intake. In: *The Chemical Senses and Nutrition*. Eds. M. R. Kare & O. Maller. 293–312. Baltimore: Johns Hopkins Press.

GROSSMAN, S. P. (1968). Hypothalamic and limbic influences on food intake. *Fed. Proc.*, **27**, 1349–1360.

HABERICH, F. J. (1968). Osmoreception in the portal circulation. *Fed. Proc.*, **27**, 1137–1141.

HAIGHTON, J. (1797). See Green, 1966a.

HALÁSZ, B. (1968). The role of the hypothalamic hypophysiotrophic area in the control of growth hormone secretion. *Proc. 1st Int. Sympos. Growth Hormone*. Excerpta Medica Int. Cong. Series. No. 158, 204–210.

HALÁSZ, B. (1969). The endocrine effects of isolation of the hypothalamus from the rest of the brain. In: *Frontiers in Neuroendocrinology, 1969*. Eds. W. F. Ganong & L. Martini. 307–342. New York: Oxford University Press.

HALÁSZ, B., FLORSHEIM, W. H., CORCORRAN, N. L., & GORSKI, R. A. (1967). Thyrotrophic hormone secretion in rats after partial or total interruption of neural afferents to the medial basal hypothalamus. *Endocrinology*, **80**, 1075–1082.

HALÁSZ, B. & PUPP, L. (1965). Hormone secretion of the anterior pituitary gland after physical interruption of all nervous pathways to the hypophysiotrophic area. *Endocrinology*, **77**, 553–562.

HALÁSZ, B., SLUSHER, M. A., & GORSKI, R. A. (1967). Adrenocorticotrophic hormone secretion in rats after partial or total deafferentation of the medial basal hypothalamus. *Neuroendocrinology*, **2**, 43–55.

HALBERG, F., HALBERG, E., BARNUM, C. P., & BITTNER, J. J. (1959). Physiologic 24-hour periodicity in human beings and mice, the lighting regimen and daily routine. In: *Photoperiodism and related phenomena in plants and animals*. 803–878. Washington: Amer. Ass. Adv. Sci.

HAMBURGH, M. (1968). An analysis of the action of thyroid hormone on development based on *in vivo* and *in vitro* studies. *Gen. Compar. Endoc.*, **10**, 198–213.

HARKER, J. E. (1958). Diurnal rhythms in the animal kingdom. *Biol. Rev.*, **33**, 1–52.

HARRIS, G. W. (1947). The innervation and actions of the neurohypophysis; an investigation using the method of remote control stimulation. *Phil. Trans. Roy. Soc.*, **B232**, 385–441.

HARRIS, G. W. (1948). Neural control of the pituitary gland. *Physiol. Rev.*, **28**, 139–179.

HARRIS, G. W. (1960). Central control of pituitary secretion. In: *Neurophysiology*. Ed. H. W. Magoun. **2**, 1007–1038. Washington: Amer. Physiol. Soc.

HARRIS, G. W. (1964). Sex hormones, brain development and brain function. *Endocrinology*, **75**, 627–648.

HARRIS, G. W. & NAFTOLIN, F. (1970). The hypothalamus and control of ovulation. *Brit. med. Bull.*, **26**, 3–9.

HARRIS, G. W. & SHERRATT, R. M. (1969). The action of chlormadinone acetate (6-chloro-Δ^6-dehydro-17α-acetoxyprogesterone) upon experimentally induced ovulation in the rabbit. *J. Physiol. Lond.*, **203**, 59–66.

HARRIS, G. W. & WOODS, J. W. (1958). The effect of electrical stimulation of the hypothalamus or pituitary gland on thyroid activity. *J. Physiol., Lond.*, **143**, 246–274.

HARRIS, I. (1966). The chemistry of intermediate lobe hormones. In: *The Pituitary Gland*. Eds. G. W. Harris & B. T. Donovan. **3**, 33–40. London: Butterworths.

HEAPE, W. (1905). Ovulation and degeneration of ova in the rabbit. *Proc. Roy. Soc. B.*, **76**, 260–268.

HELLER, H. (1966). The hormone content of the vertebrate hypothalamo-neurohypophysial system. *Brit. med. Bull.,* **22**, 227–231.

HELLER, H. & GINSBURG, M. (1966). Secretion, metabolism and fate of the posterior pituitary hormones. In: *The Pituitary Gland.* Eds. G. W. Harris & B. T. Donovan. **3**, 330–373. London: Butterworths.

HERBERT, J. (1969). The pineal gland and light-induced oestrus in ferrets. *J. Endocrinol.,* **43**, 625–636.

HILLIARD, J., CROXATTO, H. B., HAYWARD, J. N., & SAWYER, C. H. (1966). Norethindrone blockade of LH release to intrapituitary infusion of hypothalamic extract. *Endocrinology,* **79**, 411–419.

HILTON, S. M. (1966). The hypothalamic regulation of the cardiovascular system. *Brit. med. Bull.,* **22**, 243–248.

HIMWICH, W. A. (1962). Biochemical and neurophysiological development of the brain in the neonatal period. *Int. Rev. Neurobiol.,* **4**, 117–158.

HOAGLAND, H. (1954). Studies of brain metabolism and electrical activity in relation to adrenocortical physiology. *Recent. Prog. Horm. Res.,* **10**, 29–58.

HOFFMAN, J., KORDON, C., & BENOIT, J. (1968). Effect of different photoperiods and blinding on ovarian and testicular functions in normal and testosterone-treated rats. *Gen. Comp. Endoc.,* **10**, 109–118.

HOFFMAN, R. A. & REITER, R. J. (1965). Pineal gland: influence on gonads of male hamsters. *Science,* **148**, 1609–1611.

HOHLWEG, W. & JUNKMANN, K. (1932). Die Hormonal-Nervöse Regulierung der Funktion des Hypophysenvorderlappens. *Klin. Wschr.,* **11**, 321–323.

HOPE, D. B. & HOLLENBERG, M. D. (1968). Crystallization of complexes of neurophysins with vasopressin and oxytocin. *Proc. Roy. Soc. B.,* **170**, 37–47.

HORTON, E. W. (1969). Hypotheses on physiological roles of prostaglandins. *Physiol. Rev.,* **49**, 122–161.

HOWE, A. & THODY, A. J. (1969). The effect of hypothalamic lesions on the melanocyte-stimulating hormone content and histology of the pars intermedia of the rat pituitary gland. *J. Physiol. Lond.,* **203**, 159–171.

HUFELAND, C. W. (1797). See *Ann. N.Y. Acad. Sci.,* **117**, 3 (1964).

HUNTER, W. M., RIGAL, W. M., & SUKKAR, M. Y. (1968). Plasma growth hormone during fasting. *Proc. 1st Int. Sympos. Growth Hormone.* Excerpta Medica Congress Series, No. 158, 408–417.

IFFT, J. D. (1966). Further evidence on an 'internal' feedback from the adenohypophysis to the hypothalamus. *Neuroendocrinology,* **1**, 350–357.

IGARASHI, M. & McCANN, S. M. (1964). A hypothalamic follicle stimulating hormone-releasing factor. *Endocrinology,* **74**, 446–452.

JACOBSOHN, D. (1966). The techniques and effects of hypophysectomy, pituitary stalk section and pituitary transplantation in experimental animals. In: *The Pituitary Gland.* Eds. G. W. Harris & B. T. Donovan. **2**, 1–21. London: Butterworths.

JEWELL, P. A. & VERNEY, E. B. (1957). An experimental attempt to determine the site of the neurohypophysial osmoreceptors in the dog. *Phil. Trans. Roy. Soc.,* **B240**, 197–324.

JØRGENSEN, C. B. & LARSEN, L. O. (1963). Nature of the nervous control of pars intermedia function in amphibians: rate of functional recovery after denervation. *Gen. Comp. Endocrinol.,* **3**, 468–472.

JOST, A. (1966). Anterior pituitary function in foetal life. In: *The Pituitary Gland.* Eds. G. W. Harris & B. T. Donovan. **2**, 299–323. London: Butterworths.

JOST, A., DUPOUY, J.-P., & MONCHAMP, A. (1966). Fonction corticotrope de l'hypophyse et hypothalamus chez le foetus de Rat. *C. R. Acad. Sci. Paris,* **262**, 147–150.

JOUAN, P. (1968). Aspects récents de la physiologie et de la biochimie de la glande pinéale. II. Propriétés biologiques de la glande pinéale. *Pathol. et Biol.,* **16**, 209–255.

JUTISZ, M. (1967). Études *in vitro* sur le mécanisme d'action des facteurs hypothalamiques LRF et FRF. *Arch. d'Anat. Micros. Morphol. Expér.,* **56**, Suppl. 3–4, 505–514.

KAPPERS, J. A. (1965). Survey of the innervation of the epiphysis cerebri and the accessory pineal organs of vertebrates. *Prog. Brain Res.*, **10**, 87–151.

KAPPERS, J. A. (1967). The mammalian epiphysis cerebri as a center of neurovegetative regulation. *Acta Neuroveg. (Wien)*, **30**, 190–200.

KAWAKAMI, M. & SAITO, H. (1967). Unit activity in the hypothalamus of the cat: effect of genital stimuli, luteunizing hormone and oxytocin. *Jap. J. Physiol.*, **17**, 466–486.

KAWAKAMI, M., SETO, K., TERASAWA, E., & YOSHIDA, K. (1967). Mechanisms in the limbic system controlling reproductive functions of the ovary with special reference to the positive feedback of progestin to the hippocampus. *Prog. Brain Res.*, **27**, 69–102.

KAWAKAMI, M., SETO, K., & YOSHIDA, K. (1968). Influences of the limbic structure on biosynthesis of ovarian steroids in rabbits. *Jap. J. Physiol.*, **18**, 356–372.

KENDALL, J. W., GRIMM, Y., & SHIMSHAK, G. (1969). Relation of cerebrospinal fluid circulation to the ACTH-suppressing effects of corticosteroid implants in the rat brain. *Endocrinology*, **85**, 200–208.

KENDALL, J. W. & ROTH, J. G. (1969). Adrenocortical function in monkeys after forebrain removal or pituitary stalk section. *Endocrinology*, **84**, 686–691.

KENNEDY, G. C. (1966). Food intake, energy balance and growth. *Brit. med. Bull.*, **22**, 216–220.

KISSILEFF, H. R. (1969). Oropharyngeal control of prandial drinking. *J. comp. physiol. Psychol.*, **67**, 309–319.

KITAY, J. I. (1967). Possible functions of the pineal gland. In: *Neuroendocrinology*. Eds. L. Martini & W. F. Ganong. **2**, 641–664. New York: Academic Press.

KITAY, J. I. & ALTSCHULE, M. D. (1954). *The Pineal Gland*. Cambridge: Harvard University Press.

KLÜVER, H. (1944). Porphyrins, the nervous system and behavior. *J. Psychol.*, **17**, 209–227.

KOBAYASHI, F., HARA, K., & MIYAKE, T. (1968). Luteinizing hormone concentrations in pituitary and in blood plasma during the estrous cycle of the rat. *Endocrinol. Jap.*, **15**, 313–319.

KOBAYASHI, T., KOBAYASHI, T., KATO, J.. & MINAGUCHI, H. (1966). Cholinergic and adrenergic mechanisms in the female rat hypothalamus with special reference to feedback of ovarian steroid hormones. In: *Steroid Dynamics*. Eds. G. Pincus, T. Nakao, & J. F. Tait. 303–337. New York: Academic Press.

KOMISARUK, B. R. & OLDS, J. (1968). Neuronal correlates of behavior in freely moving rats. *Science*, **161**, 810–813.

KOMISARUK, B. R., McDONALD, P. G., WHITMOYER, D. I., & SAWYER, C. H. (1967). Effects of progesterone and sensory stimulation on EEG and neuronal activity in the rat. *Exper. Neurol.*, **19**, 494–507.

KRIEGER, D. T., SILVERBERG, A. I., RIZZO, F., & KRIEGER, H. P. (1968). Abolition of circadian periodicity of plasma 17-OHCS levels in the cat. *Amer. J. Physiol.*, **215**, 959–967.

KRULICH, L., DHARIWAL, A. P. S., & McCANN, S. M. (1968). Stimulatory and inhibitory effects of purified hypothalamic extracts on growth hormone release from rat pituitary *in vitro*. *Endocrinology*, **83**, 782–790.

KURACHI, K., IWATA, R., & HIROTA, K. (1968). Experimental studies on the metabolism of catecholamine in rat brain and sexual function. In: *Integrative Mechanism of Neuroendocrine System*. Ed. S. Itoh. 151–163. Hokkaido University School of Medicine.

LaBELLA, F. S. (1968). Storage and secretion of neurohypophyseal hormones. *Canad. J. Physiol. Pharm.*, **46**, 335–345.

LAWTON, I. E. & SCHWARTZ, N. B. (1968). A circadian rhythm of luteinizing hormone secretion in ovariectomized rats. *Amer. J. Physiol.*, **214**, 213–217.

LEACH, C. S. & LIPSCOMB, H. S. (1969). Adrenal cortical control of adrenal medullary function. *Proc. Soc. exper. Biol. Med.*, **130**, 448–451.

LERNER, A. B. (1966). Possible physiological function of intermediate lobe hormones in mammals. In: *The Pituitary Gland*. Eds. G. W. Harris & B. T. Donovan. **3**, 59–61. London: Butterworths.

LERNER, A. B., CASE, J. D., TAKAHASHI, Y., LEE, T. H., & MORI, W. (1958). Isolation of melatonin, the pineal gland factor that lightens melanocytes. *J. Amer. Chem. Soc.*, **80**, 2587.

LEUNG, P. M.-B. & ROGERS, Q. R. (1969). Food intake: regulation by plasma amino acid pattern. *Life Sci.*, **8**, 1–9.

LEVI, L. (1968). Sympatho-adrenomedullary and related biochemical reactions during experimentally induced emotional stress. In: *Endocrinology and Human Behaviour*. Ed. R. P. Michael. 200–219. London: Oxford University Press.

LEVIN, E. & SCICLI, G. (1969). Brain barrier phenomena. *Brain Res.*, **13**, 1–12.

LEVINE, S. (1967). The influence of gonadal hormones in infancy on adult behavior. In: *Symposium on Reproduction*. Ed. K. Lissák. 229–241. Budapest: Akademiai Kiadó.

LEVINE, S. & MULLINS, R. F. (1966). Hormonal influences on brain organization in infant rats. *Science*, **152**, 1585–1592.

LICHTENSTEIGER, W. (1969). Cyclic variations of catecholamine content in hypothalamic nerve cells during the estrous cycle of the rat, with a concomitant study of the substantia nigra. *J. Pharmacol. exper. Therap.*, **165**, 204–215.

LINCOLN, D. W. (1969). Effects of progesterone on the electrical activity of the forebrain. *J. Endocrinol.*, **45**, 585–596.

LINDSLEY, D. B., WENDT, R. H., LINDSLEY, D. F., FOX, S. S., HOWELL, J., & ADEY, W. R. (1964). Diurnal activity, behavior and EEG responses in visually deprived monkeys. *Ann. N.Y. Acad. Sci.*, **117**, 564–587.

LINZELL, J. L. (1963). Some effects of denervating and transplanting mammary glands. *Quart. J. exper. Physiol.*, **48**, 34–60.

LISS, L. (1965/66). The perivascular structures of the human hypothalamus. *Neuroendocrinology*, **1**, 166–177.

LISTER, M. (1675). See Green, 1966a.

LOSTROH, A. J. (1969). Regulation by FSH and ICSH (LH) of reproductive function in the immature male rat. *Endocrinology*, **85**, 438–445.

LYONS, W. R. & DIXON, J. S. (1966). The physiology and chemistry of the mammotrophic hormone. In: *The Pituitary Gland*. Eds. G. W. Harris & B. T. Donovan. **1**, 527–581.

LYONS, W. R., LI, C. H., AHMAD, N., & RICE-WRAY, E. (1968). Mammotrophic effects of human hypophysial growth hormone preparations in animals and man. *Proc. 1st Int. Sympos. Growth Hormone*. Excerpta Medica Congress Series, No. 158. 349–363.

van MAANEN, J. H. & SMELIK, P. G. (1968). Induction of pseudopregnancy in rats following local depletion of monoamines in the median eminence of the hypothalamus. *Neuroendocrinology*, **3**, 177–186.

MACHLIN, L. J., TAKAHASHI, Y., HORINO, M., HERTELENDY, F., GORDON, R. S., & KIPNIS, D. (1968). Regulation of growth hormone in non-primate species. *Proc. 1st Int. Sympos. Growth Hormone*. Excerpta Medica Int. Cong. Series, No. 158, 292–305.

MACLEAN, P. D. (1962). New findings relevant to the evolution of psychosexual functions of the brain. *J. Nerv. Ment. Dis.*, **135**, 289–301.

MALMEJAC, J. (1964). Activity of the adrenal medulla and its regulation. *Physiol. Rev.*, **44**, 186–218.

MANDESLO, (1658). See Stricker, 1863.

MANGILI, G., MOTTA, M., & MARTINI, L. (1966). Control of adrenocorticotrophic hormone secretion. In: *Neuroendocrinology*. Eds. L. Martini & W. F. Ganong. **1**, 297–370. New York: Academic Press.

MARLEY, E. & PATON, W. D. M. (1961). The output of sympathetic amines from the cat's adrenal gland in response to splanchnic nerve activity. *J. Physiol.*, **155**, 1–27.

MARLEY, E. & PROUT, G. I. (1965). Physiology and pharmacology of the splanchnic-adrenal medullary junction. *J. Physiol.*, **180**, 483–513.

195

MARSHALL, F. H. A. (1936). Sexual periodicity and the causes which determine it. *Philos. Trans. Roy. Soc.*, **B226**, 423–456.

MARTINI, L. (1966). Neurohypophysis and anterior pituitary activity. In: *The Pituitary Gland.* Eds. G. W. Harris & B. T. Donovan. **3**, 535–577. London: Butterworths.

MARTINI, L., FRASCHINI, F., & MOTTA, M. (1968). Neural control of anterior pituitary functions. *Recent Progr. Horm. Res.*, **24**, 439–485.

MASON, J. W. (1968a). A review of psychoendocrine research on the sympathetic-adrenal medullary system. *Psychosom. Med.*, **30**, 631–653.

MASON, J. W. (1968b). A review of psychoendocrine research on the pituitary-adrenal cortical system. *Psychosom. Med.*, **30**, 576–607.

MASON, J. W. (1968c). Organization of the multiple endocrine responses to avoidance in the monkey. *Psychosom. Med.*, **30**, 774–790.

MASON, J. W., MANGAN, G. Jr., BRADY, J. V., CONRAD, D., & RIOCH, D. McK. (1961). Concurrent plasma epinephrine, norepinephrine and 17-hydroxycorticosteroid levels during conditioned emotional disturbances in monkeys. *Psychosom. Med.*, **23**, 344–353.

MAYER, J. & AREES, E. A. (1968). Ventromedial glucoreceptor system. *Fed. Proc.*, **27**, 1345–1348.

MAYER, J. & THOMAS, D. W. (1967). Regulation of food intake and obesity. *Science*, **156**, 328–337.

McCANN, S. M. & PORTER, J. C. (1969). Hypothalamic pituitary stimulating and inhibiting hormones. *Physiol. Rev.*, **49**, 240–284.

McCORMACK, C. E. & MEYER, R. K. (1968). Evidence for the release of ovulating hormone in PMS-treated immature rats. *Proc. Soc. exper. Biol.*, **128**, 18–23.

McHUGH, P. R., BLACK, W. C., & MASON, J. W. (1966). Some hormonal responses to electrical self-stimulation in the *Macaca mulatta*. *Amer. J. Physiol.*, **210**, 109–113.

McKENZIE, J. M. (1968). Humoral factors in the pathogenesis of Graves' disease. *Physiol. Rev.*, **48**, 252–310.

McKENZIE, J. M. (1969). The pathogenesis of hyperthyroidism. *Triangle*, **9**, 2–8.

MEITES, J. (1967). Control of prolactin secretion. *Arch. Anat. Micros. Morphol.*, **56**, Suppl. 3–4, 516–529.

MESS, B. & MARTINI, L. (1968). The central nervous system and the secretion of anterior pituitary trophic hormones. In: *Recent Advances in Endocrinology.* 8th Edn. Ed. V. H. T. James. 1–49. London: Churchill.

MICHAEL, R. P. (1965). Oestrogens in the central nervous system. *Brit. med. Bull.*, **21**, 87–90.

MICHAEL, R. P. & KEVERNE, E. B. (1968). Pheromones in the communication of sexual status in primates. *Nature, Lond.*, **218**, 746–749.

MILLS, J. N. (1966). Human circadian rhythms. *Physiol. Rev.*, **46**, 128–171.

MITTLER, J. C. & MEITES, J. (1964). *In vitro* stimulation of pituitary follicle-stimulating-hormone release by hypothalamic extract. *Proc. Soc. exper. Biol. Med.*, **117**, 309–313.

MITTLER, J. C., REDDING, T. W., & SCHALLY, A. V. (1969). Stimulation of thyrotropin (TSH) secretion by TSH-releasing factor (TRF) in organ cultures of anterior pituitary. *Proc. Soc. exper. Biol. Med.*, **130**, 406–409.

MOGENSEN, G. J. & STEVENSON, J. A. F. (1967). Drinking induced by electrical stimulation of the lateral hypothalamus. *Exper. Neurol.*, **17**, 119–127.

MOORE, R. Y., HELLER, A., WURTMAN, R. J., & AXELROD, J. (1967). Visual pathway mediating pineal response to environmental light. *Science*, **155**, 220–223.

MOTTA, M., FRASCHINI, F., GIULIANI, G., & MARTINI, L. (1968). The central nervous system, estrogen and puberty. *Endocrinology*, **83**, 1101–1107.

MOTTA, M., FRASCHINI, F., & MARTINI, L. (1967). Endocrine effects of pineal gland and of melatonin. *Proc. Soc. exper. Biol. Med.*, **126**, 431–435.

MOTTA, M., FRASCHINI, F., & MARTINI, L. (1969). 'Short' feedback mechanisms in the control of anterior pituitary function. In: *Frontiers in Neuroendocrinology, 1969.* Eds. W. F. Ganong & L. Martini. 211–253. New York: Oxford University Press.

MÜLLER, E. E., ARIMURA, A., SAWANO, S., SAITO, T., & SCHALLY, A. V. (1967).

Growth hormone-releasing activity in the hypothalamus and plasma of rats subjected to stress. *Proc. Soc. exper. Biol. Med.*, **125**, 874–878.

MÜLLER, E. E. & PECILE, A. (1968). Studies on the neural control of growth hormone secretion. *Proc. 1st Int. Sympos. Growth Hormone.* Excerpta Medica Congress Series, No. 158, 253–266.

MULROW, P. J. (1967). Metabolic effects of adrenal mineralocorticoid hormones. In: *The Adrenal Cortex.* Ed. A. B. Eisenstein. 293–313. London: Churchill.

MURPHY, J. T. & RENAUD, L. P. (1969). Mechanisms of inhibition in the ventromedial nucleus of the hypothalamus. *J. Neurophysiol.,* **32**, 85–102.

MYERS, R. D. & SHARPE, L. G. (1968). Chemical activation of ingestive and other hypothalamic regulatory mechanisms. *Physiol. Behav.,* **3**, 987–995.

NAUTA, W. J. H. (1963). Central nervous organization and the endocrine motor system. In: *Advances in Neuroendocrinology.* Ed. A. V. Nalbandov. 5–21. Urbana: University of Illinois Press.

NEGRO-VILAR, A., DICKERMAN, E., & MEITES, J. (1968). Removal of plasma FSH-RF activity in hypophysectomized rats by testosterone propionate or reserpine. *Endocrinology,* **83**, 1349–1352.

NIBBELINK, D. W. (1961). Paraventricular nuclei, neurohypophysis and parturition. *Amer. J. Physiol.,* **200**, 1229–1232.

NIKITOVITCH-WINER, M. & EVERETT, J. W. (1958). Functional restitution of pituitary grafts re-transplanted from kidney to median eminence. *Endocrinology,* **63**, 916–930.

NOVALES, R. R. (1967). Melanocyte-stimulating hormone and the intermediate lobe of the pituitary: chemistry, effects and mode of action. In: *Neuroendocrinology.* Eds. L. Martini & W. F. Ganong. **2**, 241–259. New York: Academic Press.

O'CONNOR, W. J. (1962). *Renal Function.* London: Arnold.

OLDS, J. (1956). Pleasure centers in the brain. *Scientific American.* October.

OLIVECRONA, H. (1957). Paraventricular nucleus and pituitary gland. *Acta physiol. scand.,* **40**, Suppl. 136.

OOMURA, Y., ONO, T., OOYAMA, H., & WAYNER, M. J. (1969). Glucose and osmosensitive neurones of the rat hypothalamus. *Nature, Lond.,* **222**, 282–284.

OOMURA, Y., OOYAMA, H., YAMAMOTO, T., & NAKA, F. (1967). Reciprocal relationship of the lateral and ventromedial hypothalamus in the regulation of food intake. *Physiol. Behav.,* **2**, 97–115.

ORTAVANT, R., MAULEON, P., & THIBAULT, C. (1964). Photoperiodic control of gonadal and hypophyseal activity in domestic mammals. *Ann. N.Y. Acad. Sci.,* **117**, 157–192.

PALKA, Y., RAMIREZ, V. D., & SAWYER, C. H. (1966). Distribution and biological effects of tritiated estradiol implanted into the hypothalamo-hypophysial region of female rats. *Endocrinology,* **78**, 487–499.

PALKOVITS, M., ZÁBORSZKY, L., & MAGYAR, P. (1968). Volume receptors in the diencephalon. *Acta Morphol. Hung.,* **16**, 391–401.

PAPEZ, J. W. (1937). A proposed mechanism of emotion. *Arch. Neurol. Psychiat.,* **38**, 725–743.

PARKES, A. S. (1966). *Sex, Science and Society.* Newcastle: Oriel Press.

PARRY, C. H. (1825). Quoted by Reichlin, 1966.

PASTEELS, J.-L. & ECTORS, F. (1968). Sensibilité de l'hypothalamus antérieur à la progestérone et à la médroxyprogestérone. *Ann. d'Endocrinol.,* **29**, 663–678.

PECILE, A. & MÜLLER, E. E. (1966). Control of growth hormone secretion. In: *Neuroendocrinology.* Eds. L. Martini & W. F. Ganong. **1**, 537–564. New York: Academic Press.

PECILE, A., MÜLLER, E., FALCONI, G., & MARTINI, L. (1965). Growth hormone releasing activity of hypothalamic extracts at different ages. *Endocrinology,* **77**, 241–246.

PFAFF, D. W. (1968). Uptake of ^{3}H-estradiol by the female rat brain. An autoradiographic study. *Endocrinology,* **82**, 1149–1155.

PICKFORD, M. (1947). The action of acetylcholine in the supraoptic nucleus of the chloralosed dog. *J. Physiol. Lond.*, **106**, 264–270.

PICKFORD, M. (1966a). Neurohypophysis and vascular reactions. In: *The Pituitary Gland.* Eds. G. W. Harris & B. T. Donovan. **3**, 399–413. London: Butterworths.

PICKFORD, M. (1966b). Neurohypophysis and kidney function. In: *The Pituitary Gland.* Eds. G. W. Harris & B. T. Donovan. **3**, 374–398. London: Butterworths.

PLINY. Quoted by Critchlow & Bar Sela (1967).

POPA, G. & FIELDING, U. (1930). A portal circulation from the pituitary to the hypothalamic region. *J. Anat.*, **65**, 88–91.

PORTER, J. C. & JONES, J. C. (1956). Effect of plasma from hypophyseal-portal vessel blood on adrenal ascorbic acid. *Endocrinology*, **58**, 62–67.

PURVES, H. D. & SIRETT, N. E. (1967). Corticotrophin secretion by ectopic pituitary glands. *Endocrinology*, **80**, 962–968.

PURVES, H. D., SIRETT, N. E., & GRIESBACH, W. E. (1965/66). Thyrotrophic hormone secretion from pituitary transplants in hypophysectomized rats. *Neuroendocrinology*, **1**, 276–292.

RABIN, B. M. & SMITH, C. J. (1968). Behavioral comparison of the effectiveness of irritative and non-irritative lesions in producing hypothalamic hyperphagia. *Physiol. Behav.*, **3**, 417–420.

RABINOWITZ, D., MERIMEE, T. J., NELSON, J. K., SCHULTZ, R. B., & BURGESS, J. A. (1968). The influence of proteins and aminoacids on growth hormone release in man. *Proc. 1st Int. Sympos. Growth Hormone.* Excerpta Medica Int. Congress Series, No. 158, 105–115.

RAISMAN, G. (1966). Neural connexions of the hypothalamus. *Brit. med. Bull.*, **22**, 197–201.

RAISMAN, G. (1969). In: *Integration of Endocrine and Non-Endocrine Mechanisms in the Hypothalamus.* Ed. L. Martini. London: Academic Press, in press.

REICHLIN, S. (1960). Growth and the hypothalamus. *Endocrinology*, **67**, 760–773.

REICHLIN, S. (1964). Function of the hypothalamus in regulation of pituitary-thyroid activity. In: *Brain-Thyroid Relationships.* Eds. M. P. Cameron & M. O'Connor. Ciba Foundation Study Group. No. 18. 17–32. London: Churchill.

REICHLIN, S. (1966). Control of thyrotropic hormone secretion. In: *Neuroendocrinology.* Eds. L. Martini & W. F. Ganong. **1**, 445–536. New York: Academic Press.

REITER, R. J. (1968). Morphological studies on the reproductive organs of blinded male hamsters and the effects of pinealectomy or superior cervical ganglionectomy. *Anat. Rec.*, **160**, 13–24.

RICHTER, C. P. (1965). *Biological Clocks in Medicine and Psychiatry.* Springfield: Thomas.

RINNE, U. K. & ARSTILA, A. U. (1965/66). Ultrastructure of the neurovascular link between the hypothalamus and anterior pituitary gland in the median eminence of the rat. *Neuroendocrinology*, **1**, 214–227.

ROBERTS, J. S. & SHARE, L. (1968). Oxytocin in plasma of pregnant, lactating and cycling ewes during vaginal stimulation. *Endocrinology*, **83**. 272–278.

ROCHA E SILVA, M. Jr., & ROSENBERG, M. (1969). The release of vasopressin in response to haemorrhage and its role in the mechanism of blood pressure regulation. *J. Physiol. Lond.*, **202**, 535–557.

RODGER, N. W., BECK, J. C., BURGUS, R., & GUILLEMIN, R. (1969). Variability of response in the bioassay for a hypothalamic somatotrophin releasing factor based on rat pituitary growth hormone content. *Endocrinology*, **84**, 1373–1383.

ROTHBALLER, A. B. (1966). Pathways of secretion and regulation of posterior pituitary factors. *Proc. Assoc. Res. Nerv. Ment. Dis.*, **43**, 86–131.

ROTHCHILD, I. (1965). Interrelations between progesterone and the ovary, pituitary, and central nervous system in the control of ovulation and the regulation of progesterone secretion. *Vitam. Horm.*, **23**, 209–327.

ROTHCHILD, I. (1967). The neurologic basis for the anovulation of the luteal phase, lactation and pregnancy. In: *Reproduction in the Female Mammal.* Eds. G. E. Lamming & E. C. Amoroso. 30–54. London: Butterworths.

ROWLAND, V. (1966). Stereotaxic techniques and the production of lesions. In: *Neuroendocrinology*. Eds. L. Martini & W. F. Ganong. **1,** 107–132. New York: Academic Press.

SAWYER, C. H. (1966). Effects of hormonal steroids on certain mechanisms in the adult brain. *Proc. 2nd Int. Cong. Hormonal Steroids*. Excerpta Medica Int. Congress Series. No. 132, 123–135.

SAWYER, C. H. & KAWAKAMI, M. (1959). Characteristics of behavioral and electro-encephalographic after-reactions to copulation and vaginal stimulation in the female rabbit. *Endocrinology*, **65,** 622–630.

SAWYER, C. H., KAWAKAMI, M., & KANEMATSU, S. (1966). Neuroendocrine aspects of reproduction. *Proc. Assoc. Res. Nerv. Ment. Dis.*, **43,** 59–84.

SAWYER, C. H., KAWAKAMI, M., MEYERSON, B., WHITMOYER, D. I., & LILLEY, J. J. (1968). Effects of ACTH, dexamethasone and asphyxia on electrical activity of the rat hypothalamus. *Brain Res.*, **10,** 213–226.

SAWYER, W. H. & MILLS, E. (1966). Control of vasopressin secretion. In: *Neuroendocrinology*. Eds. L. Martini & W. F. Ganong. **1,** 187–216. New York: Academic Press.

SAYERS, G. & SAYERS, M. A. (1947). Regulation of pituitary adrenocorticotrophic activity during the response of the rat to acute stress. *Endocrinology*, **40,** 265–273.

SAYERS, M. A., SAYERS, G., & WOODBURY, L. A. (1948). The assay of adreno-corticotrophic hormone by the adrenal ascorbic acid-depletion method. *Endocrinology*, **42,** 379–393.

SCHACHTER, S. & SINGER, J. E. (1962). Cognitive, social and physiological determinants of emotional state. *Psychol. Rev.*, **69,** 379–399.

SCHALCH, D. S. & REICHLIN, S. (1965). Stress and growth hormone release. *Proc. 1st Int. Sympos. Growth Hormone*. Excerpta Medica Int. Congress Series, No. 158, 211–225.

SCHALLY, A. V., ARIMURA, A., BOWERS, C. Y., KASTIN, A. J., SAWANO, S., & REDDING, T. W. (1968). Hypothalamic neurohormones regulating anterior pituitary function. *Recent Prog. Horm. Res.*, **24,** 497–581.

SCHALLY, A. V. & KASTIN, A. J. (1966). Purification of a bovine hypothalamic factor which elevates pituitary MSH levels in rats. *Endocrinology*, **79,** 768–772.

SCHALLY, A. V., REDDING, T. W., BOWERS, C. Y., & BARRETT, J. F. (1969). Isolation and properties of porcine thyrotropin-releasing hormone. *J. Biol. Chem.*, **244,** 4077–4088.

SCHALLY, A. V., SAWANO, S., MÜLLER, E. E., ARIMURA, A., BOWERS, C. Y., REDDING, T. W., & STEELMAN, S. L. (1968). Hypothalamic growth hormone-releasing hormone (GRH). Purification and *in vivo* and *in vitro* studies. *Proc. 1st Int. Sympos. Growth Hormone*. Excerpta Medica Int. Congress Series, No. 158, 185–203.

SCHALLY, A. V., STEELMAN, S. L., & BOWERS, C. Y. (1965). Effect of hypothalamic extracts on release of growth hormone *in vitro*. *Proc. Soc. exper. Biol. Med.*, **119,** 208–212.

SCHAPIRO, S. (1968). Some physiological, biochemical, and behavioral consequences of neonatal hormone administration: cortisol and thyroxine. *Gen. Comp. Endocrinol.*, **10,** 214–228.

SCHARRER, E. & SCHARRER, B. (1963). *Neuroendocrinology*. New York: Columbia University Press.

SCHILDKRAUT, J. J. & KETY, S. S. (1967). Biogenic amines and emotion. *Science*, **156,** 21–30.

SCHWARTZ, N. B. (1968). New concepts of gonadotropin and steroid feedback control mechanisms. In: *Textbook of Gynecologic Endocrinology*. Ed. J. J. Gold. 33–50. New York: Hoeber.

SCHWARTZ, N. B. (1969). A model for the regulation of ovulation in the rat. *Recent Prog. Horm. Res.*, **25,** 1–43.

SCHWYZER, R. & SIEBER, P. (1963). Total synthesis of adrenocorticotrophic hormone. *Nature, Lond.*, **199,** 172–174.

SELYE, H. (1936). Thymus and adrenals in the response of the organism to injuries and intoxications. *Brit. J. exper. Pathol.,* **17,** 234–248.

SELYE, H. (1950). *Stress.* Montreal: Acta.

SHARMA, K. N. (1967). Alimentary receptors and food intake regulation. In: *The Chemical Senses and Nutrition.* Eds. M. R. Kare & O. Maller. 281–291. Baltimore: Johns Hopkins Press.

SHUTE, C. C. D. & LEWIS, P. R. (1966). Cholinergic and monoaminergic pathways in the hypothalamus. *Brit. med. Bull.,* **22,** 221–226.

SINGH, D. V., PANDA, J. N., ANDERSON, R. R., & TURNER, C. W. (1967). Diurnal variation of plasma and pituitary thyrotropin (TSH) of rats. *Proc. Soc. exper. Biol. Med.,* **126,** 553–554.

SINGHAL, R. L., VALADARES, J. R. E., & SCHWARK, W. S. (1969). Inhibition by phenobarbitone of oestrogen-stimulated increases in uterine enzymes. *J. Pharm. Pharmacol.,* **21,** 194–195.

SINHA, D. & MEITES, J. (1965). Effects of thyroidectomy and thyroxine on hypothalamic concentration of 'thyrotropin releasing factor' and pituitary content of thyrotropin in rats. *Neuroendocrinology,* **1,** 4–14.

SLOPER, J. C. (1966). The experimental and cytopathological investigation of neurosecretion in the hypothalamus and pituitary. In: *The Pituitary Gland.* Eds. G. W. Harris & B. T. Donovan. **3,** 131–239. London: Butterworths.

SLUSHER, M. A. (1964). Effects of chronic hypothalamic lesions on diurnal and stress corticosteroid levels. *Amer. J. Physiol.,* **206,** 1161–1164.

SLUSHER, M. A. (1966). Effects of cortisol implants in the brain stem and ventral hippocampus on diurnal corticosteroid levels. *Exper. Brain Res.,* **1,** 184–194.

SMELIK, P. G. (1969). The regulation of ACTH secretion. *Acta Physiol. Pharm. Neerland.,* **15,** 123–135.

SMITH, E. R. & DAVIDSON, J. M. (1968). Role of estrogen in the cerebral control of puberty in female rats. *Endocrinology,* **82,** 100–108.

SMITH, P. E. (1927). Hypophysectomy and a replacement therapy in the rat. *Amer. J. Anat.,* **45,** 205–273.

SNYDER, R. L. (1968). Reproduction and population pressures. In: *Progress in Physiological Psychology.* Eds. E. Stellar & J. M. Sprague. **2,** 119–160. New York: Academic Press.

SOKOLOFF, L. (1967). Action of thyroid hormones and cerebral development. *Amer. J. Dis. Child.,* **114,** 498–505.

SPIES, H. G., STEVENS, K. R., HILLIARD, J., & SAWYER, C. H. (1969). The pituitary as a site of progesterone and chlormadinone blockade of ovulation in the rabbit. *Endocrinology,* **84,** 277–284.

STEFANO, F. J. E. & DONOSO, A. O. (1967). Norepinephrine levels in the rat hypothalamus during the estrous cycle. *Endocrinology,* **81,** 1405–1406.

STEINER, F. A., RUF, K., & AKERT, K. (1969). Steroid-sensitive neurones in rat brain: anatomical localization and responses to neurohumours and ACTH. *Brain Res.,* **12,** 74–85.

STRICKER, W. (1863). Zu der Abhandlung über geschlechtsliche Frühreife. *Würz. med. Z.,* **4,** 138–139.

STRUMWASSER, F. (1967). Neurophysiological aspects of rhythms. In: *The Neurosciences.* Ed. G. C. Quarton, T. Melnechuk, & F. O. Schmitt. 516–528. New York: Rockefeller University Press.

STUMPF, W. E. (1968). Estradiol-concentrating neurons: topography in the hypothalamus by dry-mount autoradiography. *Science,* **162,** 1001–1003.

STUNKARD, A. J. (1968). Environment and obesity: recent advances in our understanding of regulation of food intake in man. *Fed. Proc.,* **27,** 1367–1373.

SUNDSTEN, J. W. & SAWYER, C. H. (1961). Osmotic activation of neurohypophysial hormone release in rabbits with hypothalamic islands. *Exper. Neurol.,* **4,** 548–561.

SUTER, R. B. & RAWSON, K. S. (1968). Circadian activity rhythm of the deer mouse, Peromyscus: effect of deuterium oxide. *Science,* **160,** 1011–1014.

SWERDLOFF, R. S. & ODELL, W. D. (1968). Gonadotropins: Present concepts in the human. *California Medicine,* **109,** 467–485.

SZENTÁGOTHAI, J., FLERKÓ, B., MESS, B., & HALÁSZ, B. (1968). *Hypothalamic Control of the Anterior Pituitary.* 3rd Ed. Budapest: Akadémiai Kiadó.

TAYLOR, A. N. (1969). The role of the reticular activating system in the regulation of ACTH secretion. *Brain Res.,* **13,** 234–246.

TEITELBAUM, P. (1967). The biology of drive. In: *The Neurosciences.* Eds. G. C. Quarton, T. Melnechuk, & F. O. Schmitt. 557–567. New York: Rockefeller University Press.

TEITELBAUM, P., CHENG, M.-F., & ROZIN, P. (1969). Development of feeding parallels its recovery after hypothalamic damage. *J. comp. Physiol. Psychol.,* **67,** 430–441.

TERASAWA, E. & SAWYER, C. H. (1969). Electrical and electrochemical stimulation of the hypothalamo-hypophysial system with stainless steel electrodes. *Endocrinology,* **84,** 918–925.

TERASAWA, E. & TIMIRAS, P. S. (1968a). Electrophysiological study of the limbic system in the rat at onset of puberty. *Amer. J. Physiol.,* **215,** 1462–1467.

TERASAWA, E. & TIMIRAS, P. S. (1968b). Electrical activity during the estrous cycle of the rat: cyclic changes in limbic structures. *Endocrinology,* **83,** 207–216.

TIMIRAS, P. S., VERNADAKIS, A. V., & SHERWOOD, N. M. (1968). Development and plasticity of the nervous system. In: *Biology of Gestation.* Ed. N. S. Assali. **2,** 261–319. New York: Academic Press.

TINDAL, J. S. (1967). Studies on the neuroendocrine control of lactation. In: *Reproduction in the Female Mammal.* Eds. G. E. Lamming & E. C. Amoroso. 79–107. London: Butterworths.

TOMATIS, M. E. & TALEISNIK, S. (1968). Influence of reserpine on the pituitary content of melanocyte-stimulating hormone and on hypothalamic factors which affect its release. *J. Endocrinol.,* **42,** 505–512.

TRAYLOR, R. A. & BLACKBURN, J. G. (1969). Effects of temperature on the electrical activity of the hypothalamic feeding centers. *Exper. Neurol.,* **23,** 91–101.

TURNER, S. G., SECHZER, J. A., & LIEBELT, R. A. (1967). Sensitivity to electric shock after ventromedial hypothalamic lesions. *Exper. Neurol.,* **19,** 236–244.

USHER, D. R., KASPER, P., & BIRMINGHAM, M. K. (1967). Comparison of pituitary-adrenal function in rats lesioned in different areas of the limbic system and hypothalamus. *Neuroendocrinology,* **2,** 157–174.

VALTIN, H., SCHROEDER, H. A., BENIRSCHKE, K., & SOKOL, H. W. (1962). Familial hypothalamic diabetes insipidus in rats. *Nature, Lond.,* **196,** 1109–1110.

VANDER, A. J. (1967). Control of renin release. *Physiol. Rev.,* **47,** 359–382.

VELASCO, M. E. & TALEISNIK, S. (1969). Release of gonadotropins induced by amygdaloid stimulation in the rat. *Endocrinology,* **84,** 132–139.

VELASCO, M. E. & TALEISNIK, S. (1969). Effect of hippocampal stimulation on the release of gonadotropin. *Endocrinology,* **85,** 1154–1159.

VERNEY, E. B. (1947). The antidiuretic hormone and the factors which determine its release. *Proc. Roy. Soc. B.,* **135,** 25–106.

VOOGT, J. L., CLEMENS, J. A., & MEITES, J. (1969). Stimulation of pituitary FSH release in immature female rats by prolactin implant in median eminence. *Neuroendocrinology,* **4,** 157–163.

WEBB, H. M. & BROWN, F. A. (1959). Timing long-cycle physiological rhythms. *Physiol. Rev.,* **39,** 127–161.

WETZEL, M. C. (1968). Self-stimulation's anatomy: data needs. *Brain Res.,* **10,** 287–296.

WHITE, L. E. (1965). A morphologic concept of the limbic lobe. *Int. Rev. Neurobiol.,* **8,** 1–34.

de WIED, D. (1966). Opposite effects of ACTH and glucocorticosteroids on extinction of conditioned avoidance behavior. *Proc. 2nd Int. Cong. Hormonal Steroids.* Excerpta Medica Int. Congress Series, No. 132, 945–951.

de WIED, D. (1969). Effects of peptide hormones on behavior. In: *Frontiers in Neuroendocrinology, 1969.* Eds. W. F. Ganong & L. Martini. 97–140. New York: Oxford University Press.

de WIED, D., BOHUS, B., & GREVEN, H. M. (1968). Influence of pituitary and adreno-cortical hormones on conditioned avoidance behaviour in rats. In: *Endocrinology and Human Behavior*. Ed. R. P. Michael. 188–199. London: Oxford University Press.

WINGSTRAND, K. G. (1966). Microscopic anatomy, nerve supply and blood supply of the pars intermedia. In: *The Pituitary Gland*. Eds. G. W. Harris & B. T. Donovan. **3**, 1–27. London: Butterworths.

WOLF, A. V. (1950). Osmometric analysis of thirst in man and dog. *Amer. J. Physiol.*, **161**, 75–86.

WOLFSON, A. (1966). Environmental and neuroendocrine regulation of annual gonadal cycles and migratory behavior in birds. *Recent Prog. Horm. Res.*, **22**, 177–239.

WOODBURY, D. M. & VERNADAKIS, A. (1966). Effects of steroids on the central nervous system. In: *Methods in Hormone Research*. Ed. R. I. Dorfman. **5**, 1–57. New York: Academic Press.

WOODS, J. W. (1962). In discussion. *Proc. xxii Int. Cong. Physiol.*, **1**, 611–612.

WOODS, J. W. & BARD, P. (1960). Antidiuretic hormone secretion in the cat with a centrally denervated hypothalamus. *Proc. 1st Int. Cong. Endocrinol.*, 113.

WOODS, W. H., HOLLAND, R. C., & POWELL, E. W. (1969). Connections of cerebral structures functioning in neurohypophysial hormone release. *Brain Res.*, **12**, 26–46.

WURTMAN, R. J., AXELROD, J., & KELLY, D. E. (1968). *The Pineal*. New York: Academic Press.

YANKOPOULOS, N. A., DAVIS, J. O., KLIMAN, B., & PETERSON, R. E. (1959). Evidence that a humoral agent stimulates the adrenal cortex to secrete aldosterone in experimental secondary hyperaldosteronism. *J. clin. Invest.*, **38**, 1278–1289.

YATES, F. E., BRENNAN, R. D., & URQUHART, J. (1969). Adrenal glucocorticoid control system. *Fed. Proc.*, **28**, 71–83.

ZAMBRANO, D. & DE ROBERTIS, E. (1968a). The effect of castration upon the ultrastructure of the rat hypothalamus. II. Arcuate nucleus and outer zone of the median eminence. *Z. Zellforsch.*, **87**, 409–421.

ZAMBRANO, D. & DE ROBERTIS, E. (1968b). The effect of castration upon the ultrastructure of the rat hypothalamus. I. Supraoptic and paraventricular nuclei. *Z. Zellforsch.*, **86**, 487–498.

ZARROW, M. X., PHILPOTT, J. E., DENENBERG, V. H., & O'CONNOR, W. B. (1968). Localization of ^{14}C-4-corticosterone in the 2-day-old rat and a consideration of the mechanism involved in early handling. *Nature, Lond.*, **218**, 1264–1265.

ZEILMAKER, G. H. (1965). Normal and delayed pseudopregnancy in the rat. *Acta endocr. Copenhagen*, **49**, 558–566.

Index

MADE AND PRINTED IN GREAT BRITAIN BY
MORRISON & GIBB LTD., LONDON AND EDINBURGH